THE CHEMISTRY AND BIOLOGY OF VIRUSES

HEINZ FRAENKEL-CONRAT
DEPARTMENT OF MOLECULAR BIOLOGY AND VIRUS LABORATORY
UNIVERSITY OF CALIFORNIA, BERKELEY, CALIFORNIA

ACADEMIC PRESS New York and London

1969

COPYRIGHT © 1969, BY ACADEMIC PRESS, INC.
ALL RIGHTS RESERVED
NO PART OF THIS BOOK MAY BE REPRODUCED IN ANY FORM,
BY PHOTOSTAT, MICROFILM, OR ANY OTHER MEANS, WITHOUT
WRITTEN PERMISSION FROM THE PUBLISHERS.

ACADEMIC PRESS, INC.
111 Fifth Avenue, New York, New York 10003

United Kingdom Edition published by
ACADEMIC PRESS, INC. (LONDON) LTD.
Berkeley Square House, London W1X6BA

LIBRARY OF CONGRESS CATALOG CARD NUMBER: 77-84244

268442

PRINTED IN THE UNITED STATES OF AMERICA

PREFACE

This book is intended to introduce students in the areas of biology, biochemistry, bacteriology, and the premedical sciences to the principal facts now known about the structure and function of viruses. It is also intended to help research workers in these general areas to broaden their specialization. The material is handled in the most elementary possible manner, avoiding higher mathematics and physics.

The book covers the chemistry of bacterial, plant, and animal viruses. The aspects discussed are isolation and preparation of viruses and their components, analysis and structure of the components, modification and mutagenesis, reconstitution of viruses from components, replication of viruses, and the biology of tumor virus and temperate phages.

To enable the student or researcher to delve more deeply into his particular area of interest, many more references are given than is customary for a textbook. In order to avoid interruption of text by the quotation of many names, the references are indicated by number only. The complete references, with article titles, are arranged in a numbered alphabetical list at the end of the book. The choice of references is obviously arbitrary. If the list contained 6000 instead of the actual 600 references it would still be incomplete and would give cause for justified complaint. It is intended only as a guide to the reader in his further search. About forty-five percent of the papers quoted may be regarded as samples of the classics (1798–1966), about fifty percent as the latest to come to the author's attention (1967–1969), and five percent as included on the basis of personal bias.

The thoroughness of coverage of different aspects of the field also varies, and the choice reflects the author's preferences. Again, it is hoped that the Bibliography and the Subject and Author Indexes will help the reader fill the gaps.

The subject matter of this book continues to grow at a rapid rate. Most statements could or should be amplified (or modified) within a few days or weeks after they are made, and new statements added. Thus, such a work might never materialize. Alternatively, it can be written and published within a minimum time span, and advances made in the course of that period only noted by added references. This more decisive course has been selected, the book hopefully taking no longer that a human pregnancy from conception to delivery. This concentrated effort was made possible by the secretarial assistance of Mrs. M. Hicks at Berkeley, and by the capable staff of Academic Press.

This work contains a questions section which is concerned with the main points made in each chapter. These may enable the reader (or instructor) to ascertain to what extent communication between author and reader has been accomplished successfully.

HEINZ FRAENKEL-CONRAT

Berkeley, California
April, 1969

CONTENTS

CHAPTER I

THE RECOGNITION OF VIRUSES

At the beginning of the last century, when Jenner was trying to convince his colleagues in England of the value of his smallpox vaccine, the intrinsic differences between various harmful agents, all of them frequently called by the Latin word, virus, were not yet understood. Only with the realization of the nature of bacteria in the middle of the nineteenth century did the principal difference between bacteria and the various poisons and toxins become clear. The poisons and toxins were then recognized to be harmful substances which did not increase in living tissue or spread from one organism to another. They ranged in chemical complexity from carbon monoxide to large protein molecules. Bacteria, on the other hand, were recognized as microorganisms which grow and multiply in body fluids and cells. When methods were developed to maintain bacteria in or on nutrient media *in vitro,* this feature became one of the most important criteria for their definition. Bacteria were found to be responsible for a host of infectious diseases, hopefully for all, of animals and man. Many of the more virulent bacteria were found to produce highly toxic proteins.

Whenever a science has advanced to a new level of understanding, its practitioners tend to establish a dogma and by so doing they usually retard further advances. Since communicable diseases were proclaimed to be generally caused by microorganisms, failure to find and grow the organism causing a given disease was likely to be regarded as an indication of poor technique. Scientific or philosophical nonconformists are needed at such times to question the dogma. Beijerinck, at Leiden, was such a man. Although he failed to convince Iwanowski, at St. Petersburg, that they both had demonstrated a new phenomenon in transmitting the tobacco mosaic disease by means of bacteria-free extracts, he proclaimed a new concept in 1898 from which slowly grew an understanding of the nature of what are now called viruses. In Table 1.1 are delineated the steps in the development of virology over the subsequent seven decades.

Among the main events in this period which led virology from a descriptive to an experimental science are probably the isolation of pure viruses, starting with Stanley's work on tobacco mosaic virus; the elucidation of phage replication by Delbrück and collaborators; the identification of DNA and RNA as the active principles of viruses by Hershey and Chase, by Gierer and Schramm, and by Fraenkel-Conrat; the reconstitution of viruses and mixed viruses by Fraenkel-Conrat with Williams and Singer; the renaturation of virus proteins by Anderer; the amino acid sequence analysis of TMV in the laboratories of Schramm and Fraenkel-Conrat; and the *in vitro* translation of phage RNA and replication of RNA and DNA in Nathans', Spiegelmann's, and Kornberg's laboratories.

The first definitions of the new type of agents, the viruses, were presented entirely in negative terms: They were not retained by bacterial filters, not visible in the microscope, not cultivatable on any nutrient media *in vitro*. The first two of these properties are a consequence of viruses being generally smaller than bacteria, although it is now known that the microscopically detected elementary bodies associated with smallpox actually represent the smallpox virus. Furthermore all viruses can now be made visually apparent by electron microscopy. Concerning filtrability, refinements in the preparation of membranes of desired porosity enabled Elford in the 1930s to differentiate viruses on the basis of their size.

Only the third distinguishing characteristic of viruses listed above, that they "increase and multiply" only in living tissue or cells, is a truly fundamental issue that supplies a real basis for their definition. One other point of discrimination between viruses and living orga-

TABLE 1.1
DEVELOPMENT OF MOLECULAR VIROLOGY

		References
1798	Development of smallpox vaccination	Jenner (224)
1885	Development of rabies vaccination	Pasteur (332)
1898	Realization of the existence of a new type of infective agent	
	in mosaic diseased tobacco extracts	Beijerinck (32)
	in foot and mouth diseased tissue	Loeffler and Frosch (274)
1911	Discovery of a chicken tumor virus	Rous (364)
1915	Discovery of bacteriophages	Twort (475)
1917	Discovery of bacteriophages	d'Herélle (179)
1927–1931	Isolation of impure tobacco mosaic virus (TMV)	Vinson and Petre (487, 488)
1934	Isolation of bacteriophages	Schlesinger (387)
1935	Isolation of TMV in paracrystalline form	Stanley (431)
1937	Recognition of the nucleoprotein nature of TMV	Bawden and Pirie (28)
1938	Comparative study of virus particle sizes	Elford (101)
1940	Initiation of quantitative phage research	Delbrück (79)
1950	Induction of prophage of lysogenic phages	Lwoff *et al.* (278)
1952	Indications that phage DNA, not coat, is infective	Hershey and Chase (183)
1952	Discovery of general transduction	Zinder and Lederberg (533)
1955	Crystallization of poliomyelitis virus	Schaffer and Schwerdt (385)
1955–1957	Reconstitution of infective TMV	Fraenkel-Conrat, Williams and Singer (128, 125)
1956	Demonstration of infectivity of TMV RNA	Gierer and Schramm (147) Fraenkel-Conrat (115)
1960	Amino acid sequence of TMV protein	Anderer *et al.* (9) Tsugita *et al.* (474)
1962	*In vitro* translation of phage RNA to phage protein	Nathans *et al.* (318)
1965	*In vitro* replication of phage RNA	Haruna and Spiegelmann (170, 428)
1967	*In vitro* replication of phage DNA	Goulian *et al.* (153)

nisms that was recognized more recently is that viruses contain but one type of nucleic acid, DNA *or* RNA, while all organisms contain both.

The nature of viruses. The definition of viruses is facilitated by a prior consideration of the nature of life. Life may best be defined as a state of biophysical and biochemical organization which enables (a) a multitude of enzymes to cooperatively produce and utilize chemical energy for the synthesis of macromolecules, including themselves, and the nucleic acids, and (b) the nucleic acids to carry encoded in their structures the information, in the form of a template, for production of those macromolecules. The coordination of the activities of metabolically active proteins and information-carrying nucleic acids assures the functioning and the continuity of the living process. However, the ability of life to develop from the primordial glob to the magnificent heights of self-destructive *Homo sapiens* required the possibility of minor errors leading to mutations. This facet of life, the ability to experiment and to test something new, is now recognized as one of the most important chemical properties of the nucleic acids.

The earlier description of viruses as not being cultivatable outside the living cell indicates that they are not fully alive; that they lack the metabolic capabilities, item (a) above, of living systems. Viruses are now defined as follows: Particles made up of one or several molecules of DNA or RNA, and usually but not necessarily covered by protein, which are able to transmit their nucleic acid from one host cell to another and to use the host's enzyme apparatus to achieve their intracellular replication by superimposing their information on that of the host cell; or occasionally to integrate their genome in reversible manner in that of the host and thereby to become cryptic or to transform the character of the host cell.

The parasitic nature of viruses and their more or less complete integration into normal cellular activity represent a major obstacle in the chemotherapeutic control of virus diseases. On the other hand, the relative simplicity of extracellular viruses (the dormant phase) and the complexity of their intracellular fate and replication (the vegetative phase) have placed viruses among the most useful tools in the study of all phenomena related to replication, information transfer, mutation, and many other aspects of molecular biology. The great interest in viruses now shared by biochemists, biophysicists, geneticists, bacteriologists, and general biologists is not primarily focused on problems of their taxonomic classification, pathogenicity, or therapeutic control, but on the principles of their structure and function. This book is therefore so organized as to stress the fundamental similarities and differences in the structure and replication of different types of virus particles, irrespective of whether they infect plants, bacteria, or animals including man, and of whether they are pathogenic or not. The reader who is interested in specific studies concerned with a particular virus will find the Subject Index a useful guide to the discussion of that virus in the various chapters.

CHAPTER II

THE SYMPTOMS OF VIRAL INFECTION, AND VIRUS ASSAYS

The effects of virus infection on host organisms, tissues, or cells can range from nondetectability to severe disease symptoms and death. Plants and animals may become generally (systemically) affected, or they may show only a localized response at the site of application. The local lesion response to plant virus application on a leaf (197) corresponds to that produced on the chorioallantoic membrane of the chick embryo by many animal viruses, as well as to the plaques produced by these viruses on monolayers of animal cells in tissue culture (90), and by bacteriophages on bacterial lawns (102). In each case the size and appearance of the lesions resulting from a focus of infection can give important information about the virus responsible for it, and the number of such lesions reflects the number of infective particles applied to the test object. Thus the local lesion response is useful for purposes of diagnosis and quantitation. The nature of the generalized systemic response of tissues or organisms also serves frequently in identifying a virus. Furthermore, generally infected tissue yields the most advantageous starting material for the preparation of pure viruses.

This description of the principles of detection and identification of viruses is grossly oversimplified. Although it is easy to achieve these objectives for well-known viruses, the characterization of a new virus available in only trace amounts in a tissue sample or biological fluid is a difficult and often unsuccessful task. The reason is that most viruses are quite selective, growing only in one type of tissue or in a particular variant of a host, and often producing either no pathological changes or barely detectable symptoms. The inoculum may well be exhausted before a host giving either a local lesion response or a systemic reaction has been detected.

Another aspect which complicates the search for unknown viruses is the variability in the mechanism of infection. The bacteriophages of *Escherichia coli* have developed more or less sophisticated organs for piercing or bypassing the cell wall, while certain phages growing on *Bacillus subtilis,* as well as most animal viruses, seem able to enter the host cell at specific sites without special mechanisms. In contrast, most plant viruses are dependent on mechanical damage to the cell wall, usually of the hair cells, before they can enter. Such damage results frequently from physical forces (wind, abrasive rubbing), but in other instances the virus must be introduced into the cell by a particular animal vector (insects, mites, nematodes, etc.) or a fungus. With certain viruses, infection of a plant can be achieved in the laboratory only by grafting with infected tissue or through the roots. The interested reader is referred to *Methods in Virology,* Vol. I (Maramorosch and Koprowski, eds., Academic Press, New York, 1967) for several chapters on the transmission of viruses.

An important prerequisite for most biological research is the availability of one or more quantitative tests. It is therefore not surprising that the periods of great advances in knowledge of both plant and animal viruses were preceded by the development of reliable means of quantitating viral infectivity (90, 102, 197).

The principle of the plant leaf local lesion assay has been discussed above. Rather than discuss the complicated mathematics generally claimed to be necessary for the quantitative evaluation of local lesion numbers, on the basis of an enormous amount of data the reader can be assured that, in a typical assay system (tobacco mosaic virus, TMV,* applied with carborundum in 0.1 *M*

* All virus abbreviations that are used can be found as such, together with the complete name, in the Subject Index.

phosphate of pH 7.0 to *N. tabaccum*, var. *Xanthi* nc), over an approximately fiftyfold range in virus concentration, lesion numbers are directly proportional to virus concentration. This is so for a given set of plants and conditions, notwithstanding the fact that the sensitivity of groups of plants may vary by as much as an order of magnitude. Thus, depending on season, weather, nutritional state, and age of plants, as well as other unknown causes, 0.1 ml of an 0.01 μg/ml (10^{-8} g/ml) solution of TMV applied to a leaf may, in a given test, produce 50 lesions; one-tenth that concentration of virus (0.001 μg/ml), 5 lesions; and a fivefold (0.05 μg/ml), 250 lesions. Naturally such numbers are obtained only as the average of many infected leaves. To minimize the variability of the response, unknown samples or solutions are usually compared to a known concentration of the pure virus, if it is available, and both solutions are distributed over equivalent leaf areas. Often one half of a leaf is used for the unknown, and the opposite half for the known (standard) sample. The relative infectivity of the unknown can thus be determined on about 6 plants (30 half leaves for unknown and standard) with an accuracy of about $\pm 15\%$. Figure 2-1 shows examples of local lesions caused by TMV and a mutant giving smaller lesions on Xanthi tobacco.

Bacteriophage-containing solutions are similarly tested by plating them with an excess of susceptible bacteria (see Fig. 2-2). Most animal viruses can similarly be tested on tissue culture plates containing a monolayer of susceptible cells (90). Direct proportionality is generally accepted for the bacterial and animal viruses between particle count and foci of infection. Nevertheless, biological variability and other factors affect all these methods. They are not comparable to the titration methods of the chemist, and preferably they should be termed virus assays rather than virus titrations. This is particularly advisable since in most systems the absolute numbers of virus particles is not determined, but rather a number lacking a direct physical significance, namely, the number of infective units or plaque-forming units (PFU) per milliliter.

In the case of the T-even bacteriophages the ratio of particles to infective units has been shown under the most favorable conditions to be between 1 and 2 (276), but for animal and plant viruses it ranges from 10 to 10^6. This could be and has often been interpreted

FIG. 2-1. Plant virus lesions. Local lesions produced by common TMV (the large ones) and certain strains of TMV (the smaller and variable ones on the opposite half leaves) on Xanthi plants 6 days after inoculation.

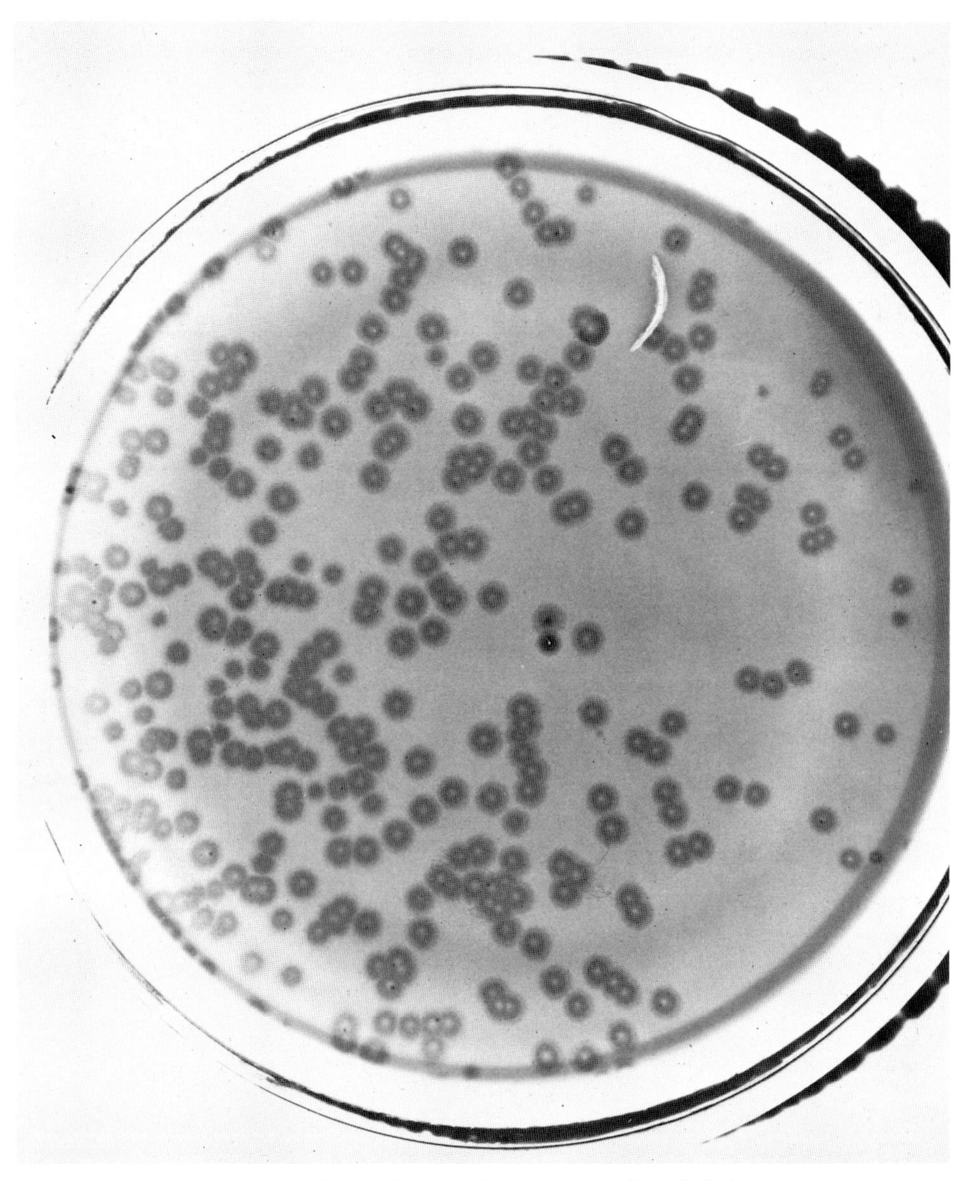

FIG. 2-2. Bacteriophage plaques. T2r plaques on *E. coli* Br+. An r mutant near top. (Courtesy of G. S. Stent.)

as signifying that most virus particles are not infectious. However, the more probable explanation is that all typical virus particles are potentially infective but that most of them happen not to be able to initiate an infection under the test conditions. The previously mentioned fact, that many coliphages possess an infection-favoring organ, but that animal and plant viruses infect only through certain cell receptor sites or wounds, suggests one type of explanation for the lesser efficiency of the latter two classes of viruses.

It must be stressed that quantitative evaluation of the infectivity in terms of virus particles is possible regardless of the ratio of physical particles to infective centers, provided that purified virus is available, the infectivity of which can be compared to that of the unknown infective material. Furthermore, it should be noted that the proportionality between concentration and lesion number that can be observed with most viruses over a certain range of concentrations is a strong indication that infection generally results from a single particle invading a cell, and that no cooperative action of several or many particles is required for infectivity. This does not preclude, however, the entry into and replication of several infective particles in one host cell. Exceptional cases in which virus infections require the entry of several particles are pointed out later (see Section VIII, E).

When plaque or local lesion assays are not available for a given plant or animal virus, various less quantitative means of estimating the concentrations of infective virus must be considered. One of these techniques determines by serial dilution the minimal concentration required to elicit the typical disease symptoms. In other instances the time necessary to produce a certain cytopathic response can be used as a measure of virus concentration.

CHAPTER III

METHODS FOR ISOLATION AND PURIFICATION OF VIRUSES

A. General Principles and Terms. Classification of Viruses

Viruses generally are built up on either of two structural principles, the helical or the isometric (see Chapter VIII), which give them either a longitudinal or a roughly spherical shape, respectively. However, threadlike viruses are also known which are not strictly helical, and helical components may be embedded in a spherical virus.

The minimal structural features for typical viruses consist of one molecule of nucleic acid integrated into a shell or coat made up of many identical protein molecules. The more complex viruses may contain several molecules of nucleic acid, as well as several proteins, multiple organelles, or heterogeneous envelopes.

A group of terms has been proposed to identify the various structural components of viruses. The monomer of the protein which forms the viral coat is called the *structural subunit.* In isometric viruses several of these subunits usually form *morphological units* of a characteristic shape discernible on electron micrographs; these are also called *capsomeres.* The orderly complex of many subunits or capsomeres with viral nucleic acid is the *nucleocapsid,* as con-

TABLE 3.1

	Virus group	*Particle*	*Dimensions, mμ*	*Particle weight, daltons × 10⁶*
	RNA Viruses			
1	Rod-shaped plant viruses	Flexible rod	11 × 540	35
		Rigid rod	18 × 300	40
		Rigid rod	26 × 75–190	—
2	Picorna viruses	Isometric	20–30 (17)[a]	4–7 (2)[a]
3	Encephalo viruses	Isometric, membrane	35–50	50
4	Rhabdo viruses	Stubby rod, membrane	70 × 175	—
5	Leuko viruses	Pleomorphic, envelope	100–120	~ 500
6	Myxo viruses	Pleomorphic, envelope	80–100	~ 300
7	Paramyxo viruses	Pleomorphic, envelope	100–200	~ 700
8	Double stranded RNA viruses	Isometric	60–90	~ 80
	DNA Viruses			
9	Small DNA phages (single stranded)	Flexible rod	5 × 800	11.3
		Isometric	30	6.2

[a] The defective satellite of TNV (STNV).

[b] The nucleocapsid of the myxoviruses and of the paramyxoviruses has a diameter of about 9 and 18 mμ, respectively.

CLASSIFICATION OF VIRUSES

Group	*Capsid*	*Examples*	*Hosts*	*Site of replication and other comments*
1	Helical Helical Helical	Potato X Tobacco mosaic Tobacco rattle, barley stripe	Plants (arthropods)	Nucleus RNA, cytoplasm protein?
2	Icosahedral	Polio, turnip yellow mosaic, f2, Qβ	Mammals, plants, bacteria	Cytoplasm, some hemagglutinin
3	Helical?	Equine encephalitis, Sindbis, yellow fever	Vertebrates, arthropods	Cytoplasm, budding, hemagglutinin
4	Helical	Vesicular stomatitis	Mammals, plants	Cytoplasm, budding, hemagglutinin
5	Helical	Rous sarcoma, mammary tumor	Rodents, birds	Oncogenic
6	Threads[b]	Influenza	Mammals	Nucleus?, budding, hemagglutinin, neuraminidase
7	Helical[b]	Mumps, measles, Newcastle disease	Vertebrates	Cytoplasm, budding, some neuraminidase
8	Icosahedral	Reo, wound tumor, rice dwarf	Mammals, plants	Cytoplasm, hemagglutinin
9	Helical (α) Dodecahedral	fd, M13 φX174, S13	Bacteria Bacteria	Extrusion Lysis

TABLE 3.1 (*Continued*)

Nucleic acid	*Virus group*	*Particle*	*Dimensions, mμ*	*Particle weight, daltons × 10⁶*
10[c]	Parvo viruses	Isometric	18–22	—
11[c]	Adeno-associated viruses	Isometric	28	—
12	Papova viruses	Isometric	30–50	40
13	Medium-sized phages	Isometric, tailed	50 (15–150)[d]	50
14	Large phages	Oblong, tailed	65 × 95 (110)[d]	300
15	Adeno viruses	Isometric	80–90	175
16	Herpes viruses	Isometric	100–150	~ 800
17	Iridescent viruses	Isometric	130–170	~1100
18	Paravaccinia viruses	Ovoid	160 × 260	~2000
19	Pox viruses	Brick-shaped, membrane	100 × 200 × 300	~3000

[c] A recent study indicates that viruses in class 9, 10, and 11 have similar dimensions and sedimentation constants, and single-stranded DNA of similar molecular weight (1.7 × 10.6) (537).

[d] Tail length (diameter 10 mμ). There are many more medium-sized phages of different morphology and tail length.

CLASSIFICATION OF VIRUSES (*Continued*)

Group	*Capsid*	*Examples*	*Hosts*	*Site of replication and other comments*
10[c]	Icosahedral	Minute murine, X14	Rodents, bacteria	Tumors, latency
11[c]	Icosahedral	AAV	Mammals	Defective
12	Icosahedral	Polyoma, papilloma, SV40	Vertebrates	Nucleus (DNA, assembly), cytoplasm protein, oncogenic, hemagglutinin
13	Icosahedral	λ, T1, etc.	Bacteria	Lysogenic and virulent
14	Icosahedral	T-even	Bacteria	Contain HMC, virulent
15	Icosahedral	Adeno type 1-31, SV20	Mammals, birds	Nucleus (DNA, assembly), cytoplasm (protein), hemagglutinin, oncogenic
16	Icosahedral	Herpes simplex, varicella	Mammals	Nucleus (DNA, assembly), cytoplasm (protein), latency
17	Icosahedral	Tipula iridescent	Insects	Cytoplasm
18	Complex	Orf	Mammals	Cytoplasm
19	Complex	Vaccinia, myxoma	Mammals, birds	Cytoplasm, hemagglutinin

trasted to the empty *capsid*, the aggregated protein shell lacking nucleic acid. The complete virus particle, which in addition to the nucleocapsid may contain additional structural proteins or envelopes, is called the *virion*.

Another point of semantics: The terms molecule and *molecular weight* are used here only for compounds held together by primary linkage. Thus a virus is never a molecule but, in the simplest cases, an aggregate of many protein molecules with one molecule of a single stranded nucleic acid. A capsomere is an aggregate of protein molecules and has a characteristic *particle weight* but not a molecular weight. However, definition of the complementary strands of double stranded DNA causes a quandary. The term double molecule, or simply molecule, is used here for this singularly specific molecular aggregate.

No rational means exists for the classification of viruses since they lack a recognizable common evolutionary relationship. As is pointed out later, some viruses may well have arisen from parasitic microorganisms through regression, while others may represent cell-derived particles on a separate evolutionary path.

Notwithstanding this probable phylogenic heterogeneity, many viruses have come to resemble one another in physical, chemical, and biological properties and thus invite grouping. The least useful system of classification appears to be the one based on the nature of the host, since many viruses replicate in several quite different hosts while others can lose the ability to grow in a particular host through a single-step mutation.

The methods of classification which seem least fraught with paradoxes and ambiguities are based on the chemical and physical characteristics of the viruses. This principal approach has been largely accepted. Only for one group of not closely related animal viruses, the arboviruses, has a name been advocated on the basis of their transmission, for they are *ar*thropod-*bo*rne. We agree with Fenner (109) that this is an unnecessary inconsistency, and that encephaloviruses is a better name for this group.

Table 3.1 lists nineteen virus groups on the basis of type and strandedness of nucleic acid, shape and dimensions of the particle, and presence of envelopes or tails, as well as capsid symmetry, with a few examples of each group. The classification of the animal viruses is that used by Fenner. The plant viruses are not classified beyond segregation of the rod-shaped, the isometric, and the double stranded RNA groups. The bacterial viruses also are segregated only according to the character of the nucleic acid and to an arbitrary classification of size groups, the T-even phages being called large, and all other phages containing double stranded DNA, medium-sized. A more important criterion is the substitution of cytosine by hydroxymethylcytosine (HMC) in the T-even phages.

This classification differs from that used by others only in combining all small isometric RNA viruses (17–30 mμ diameter) under the term *picorna* viruses (pico = small; thus small RNA), a term which was originally proposed only for small animal viruses. However, the similarities of these viruses and the majority of the plant viruses as well as the small RNA phages make it preferable to remove the unnecessary host restriction from this quite general term.

It appears that groups 2, 3, 4, and 8 contain viruses which infect animals and others which infect plants; group 10 includes bacterial and animal viruses. Viruses which replicate in both arthropods and plants or vertebrates occur in groups 1–3. New groups are likely to be added to this list. Only one group of insect-pathogenic viruses (#17) has been included because the molecular characterization of most others is as yet insufficient for classification.

B. General Methods of Purification of Viruses

The purification of viruses follows general procedures of protein isolation. To separate viruses from cellular proteins, use is made of the particulate nature of viruses, and their higher buoyant density, which is due to their nucleic acid component. However, special techniques are necessary to separate viruses from cellular nucleoprotein particles of similar composition and size.

Before purification or experimental use of a virus it is advisable to ascertain its stability in terms of loss of infectivity under various conditions. Usually viruses in infected tissue are not harmed by freezing and are thus preferentially stored and accumulated at low temperatures, but there are exceptions—viruses which are degraded by freezing. Many purified viruses (e.g., TMV) are harmed by repeated freezing and thawing. Most viruses are not harmed by high salt concentrations, but some (e.g., bromegrass mosaic virus, BMV) are degraded by molar salt and can thus not be subjected to certain purification procedures. All viruses are degraded by extremes of pH and by high temperatures, but they vary greatly in their pH stability range and temperature sensitivity. An example is the acid sensitivity of foot and mouth disease virus (FMDV), which loses infectivity below pH 6.5. On the other hand, BMV swells at neutrality and is stable only at pH 4–6. Thus pH and temperature conditions can often be selected which are harmful and denaturing to contaminating components but not to the virus to be isolated.

Even one of the gentlest procedures for the isolation of viruses, differential centrifugation, is harmful to certain viruses (e.g., influenza virus) which are sensitive to being packed by forces of about 100,000*g* at the bottom of the cell. Such viruses can be isolated successfully by placing a concentrated sucrose solution in the bottom of the tube.

On the basis of such preliminary stability tests, methods of purification can be designed to fit each particular case. With some of the more stable viruses, such as TMV, which also occur frequently in very high concentrations in the infected tissue, almost any method or combination of methods can be used.

All virus isolations start out with the homogenization, by grinding or blending, of infected cells or tissues with water or, more frequently, with a buffer which will assure a favorable final pH for the extract. When permissible, low temperature freezing facilitates the grinding of tissues. Somewhat alkaline buffers (e.g., K_2HPO_4) are often advantageous for the extraction of viruses from plant tissues to neutralize their acidic components. The extract is then separated from the coarse tissue debris by filtration through cheesecloth or paper.

Preferred procedures for the isolation of viruses, starting with such tissue extracts, are (a) differential centrifugation, (b) ammonium sulfate precipitation, (c) sucrose gradient centrifugation, (d) cesium chloride density gradient centrifugation, (e) solvent extraction and countercurrent distribution between aqueous polymer solutes, and (f) enzyme treatments (433, 434).

(a) Differential centrifugation consists in alternating cycles of high and low speed centrifugation. Centrifugation at 20,000 to 50,000 rpm for 1–3 hr sediments viruses together with other particulate material, and leaves soluble components of the extract of less than 10^6 molecular weight in the supernatant. The sediment is then dispersed in water or in a suitable buffer (this is generally done by allowing the pellet to soften rather than by vigorous stirring) and recentrifuged at 8000–10,000 rpm to remove membranous and fibrous cell components, denatured proteins, etc. Subsequent centrifugation at 30,000–50,000 rpm again brings down viruses, and two or three repetitions of this procedure, discarding low speed sediments and high speed supernatants, yields TMV and many other plant viruses in comparatively pure form.

(b) At times it is advantageous to combine this procedure with an ammonium sulfate salting-out step. For this purpose the crude virus sediment is redissolved in water and treated with gradually increasing amounts of saturated ammonium sulfate. When a precipitate appears, it is separated by centrifugation (8000 rpm), and more ammonium sulfate is added slowly to the supernatant until another precipitate forms and is centrifuged off. Infectivity tests indicate which ammonium sulfate concentration salts out the virus, and the procedure can then be repeated and refined to remove as much nonviral components as possible by ammonium sulfate below the level of virus precipitation; the supernatant is discarded after virus precipitation.

(c) A very useful technique for the purification of viruses and their components is centrifugation through increasing concentrations of sucrose or glycerol (e.g., 5–25% sucrose) (41, 42, 299). Such gradients are easily prepared, and they afford good separations of viruses or other particles of various weights and dimensions, the sucrose stabilizing the system so that the zones containing various fractions can be separated. Viruses form bands that can often be revealed by illuminating from the top with a strong light; the virus band can then be removed by means of a syringe. Alternatively, and when dealing with lower molecular weight materials which do not scatter light, or with viruses at low concentration, a hole is pierced in the bottom of the tube and the contents are slowly drained into a series of tubes and tested for the presence of macromolecules. For large scale work, zonal centrifuges with cylindrical rotors (e.g., Spinco model ZU) can be used.

(d) In contrast to sucrose gradient centrifugation, which utilizes primarily the different sedimentation rates of various components of a mixture (see Section III,C,2), isopyknic equilibrium centrifugation in salt solutions of high density (e.g., cesium chloride) allows separation of components strictly on the basis of their buoyant densities (see Section IV,C), (305, 306). This method finds application particularly in the separation of viruses of different nucleic acid content, as well as in the characterization of nucleic acids (see Section IV,C). Density gradient centrifugation has proved of great usefulness in the purification and separation of certain viruses.

(e) Since most viruses are more stable than cytoplasmic proteins and particles, controlled denaturing conditions are frequently of considerable usefulness. Besides heat and pH adjustments, homogenization and extraction with immiscible solvents (e.g., butanol, chloroform) frequently leave the virus in the aqueous phase, while denatured material aggregates in the interphase, and lipids, etc., dissolve in the organic phase (433). Repeated treatment of the aqueous phase with a suitable immiscible solvent can achieve considerable purification of the denaturation-resistant component.

A different principle is utilized by a method exploiting differences in the partition coefficients of viruses and contaminants in liquid two-phase systems (6). One of the suitable solvent systems is dextrane sulfate and polyethylene glycol, both in aqueous solution. Countercurrent distribution can be used to maximize differences in partition coefficients between viruses and contaminants.

(f) The particulate structure of viruses which are stabilized by a great number of intersubunit and intercomponent bonds makes them generally more resistant to both nucleases and proteases or peptidases than are most soluble cellular nucleic acids and proteins. Thus virus purification is frequently aided, particularly at early stages, by controlled treatment with ribonuclease and/or deoxyribonuclease, trypsin, chrymotrypsin, pronase, or papain, the enzymes feeding on the cellular nucleic acids or proteins rather than on the virus.

C. General Methods of Characterization of Viruses

1. *Electron Microscopy (and X-Ray Diffraction)*

The electron microscope is one of the most powerful tools of the virologist. The image is the result of the scattering of the electron beam by the electrons in the specimen and is thus favored by the presence of heavy atoms in the sample. The preparation of the specimen is of greatest importance. The necessary drying of the virus-containing aqueous solution makes the presence of nonvolatile salts quite undesirable. The flattening of hollow objects which occurs on drying is considerably diminished by freeze-drying the sample. Another technical advance is shadow casting, which consists in monodirectionally depositing vaporized heavy metals on the specimen and thus creating shadows behind it which reveal much of its three-dimensional shape (21, 515). The most beautiful illustration of this technique is the image of the large insect virus, tipula iridescent, shadowed from two angles and revealing its icosahedral symmetry (see Section VIII,B) (513).

More recently, staining procedures have been introduced which differ from shadowing in that they do not increase the dimensions of the objects. Negative staining has yielded particularly clear and informative images (47, 201). It consists in applying the heavy atom stain (phosphotungstic acid or uranyl acetate) under conditions of neutrality; then it does not combine with the viral components but stains the background, including all holes and crevices so that the particles are revealed in great detail. Understanding of the structure of isometric viruses was greatly furthered by this technique.

Electron microscopy serves also as a quantitative tool in that it allows the counting of typical virus particles. To that end the volume of solution in which particles are to be counted must be known. This is ascertained by adding a definite number of indicator particles to the solution. From the ratio of TMV rods to polystyrene latex spheres in a number of microdrops the absolute number of TMV rods per milliliter can be calculated if the number of indicator spheres per milliliter is known (see Fig. 3-1) (516).

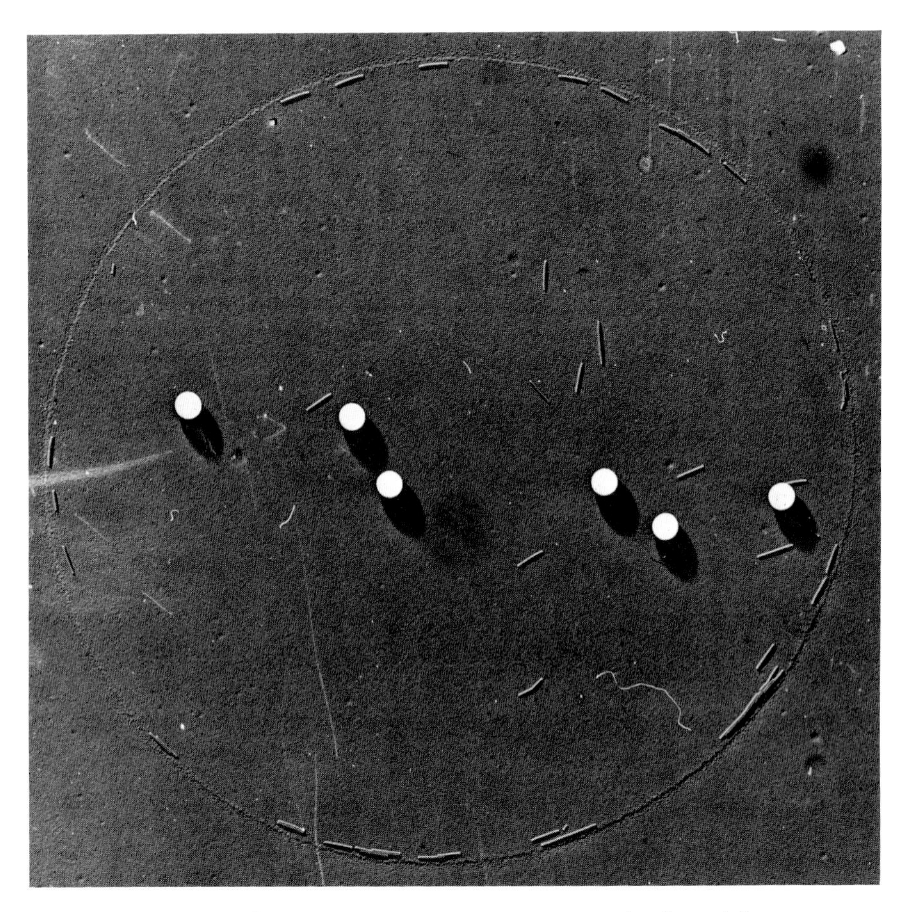

FIG. 3-1. TMV drop pattern. Virus concentration can be derived from ratio of TMV rods to polystyrene latex spheres (260 mμ diameter). (Courtesy of R. C. Williams and J. Toby.)

X-ray diffraction analysis of virus crystals and pseudocrystals has supplied quantitative data on their internal structure which have extended and complemented the electron microscopic information. Thus the subunit and capsomere topography has been established for many viruses (60, 251, 112). Because the equipment, techniques, and interpretations needed for the X-ray work are too complex and costly for general virological or biochemical laboratories, a more detailed exposition of this subject is not given here. The same is true for *light scattering* as a method for particle or molecular weight determination. The reader is referred to excellent reviews of these techniques as applied to viruses (112, 378).

2. *Ultracentrifugation**

Ever since centrifuges were developed that reached speeds of 40,000 rpm or more, the sedimentation behavior of viruses has been studied in them to the advantage of both the virologist and the physicochemist (383, 384, 288). Since the sedimentation rate is often very much affected by the concentration of the macromolecule (or particle), low concentrations assuring minimal interaction are advantageous, and the development of UV optics has proved a great boon for the study and detection of viruses and their components in dilute solution in the ultracentrifuge cell.

Sedimentation velocity is the most frequently determined parameter; it allows the calculation of the sedimentation constant s. To derive molecular (or particle) weight data from such analyses, the diffusion constant D and the partial specific volume $\bar{v}$ must also be ascertained. The latter is often derived from the composition, a procedure which permits much error and probably accounts for the spread in molecular weight data obtained for some viruses.

The calculation of the sedimentation coefficient from the photographic plate of a sedimenting boundary uses the equation $s = (1/\omega^2 r)dr/dt$ in which r is the distance of the sedimenting boundary, in centimeters, from the center of rotation, t is the time, in seconds, and ω is the angular velocity, in radians per second. The sedimentation coefficient s is expressed in Svedberg units ($1\ S = 10^{-13}$ sec). Since sedimentation runs are performed under various conditions, the observed sedimentation constant (s_{obs}) is usually expressed in terms of standard conditions, defined as 20° in water ($s_{20,w}$). For necessary equations and further detail concerning ultracentrifugation other sources should be consulted (e.g. 236).

* For a discussion of density gradient centrifugation see Sections III,B and IV,C.

Determination of the sedimentation constant with UV optics represents one of the quickest and most economical means of characterizing macromolecules and particles. About 1 absorbancy unit (0.2 mg of a typical virus) is needed and can be recovered from the cell, and the time required for centrifugation is of the order of 1 hr. However, as stated, the sedimentation rate is greatly dependent on the conformation of the macromolecules. Thus no conclusions concerning molecular weights can be derived from sedimentation constants alone. Only under identical conditions in regard to ionic medium, temperature, pretreatment, etc., can identical sedimentation values be expected for a given macromolecule. An empirical formula for deriving the molecular weight of viral RNA's lacking a specific conformation has been proposed on the basis of studies with TMV RNA ($M = 1550\ S^{2.1}$) (429).

In contrast to the determination of sedimentation rates the *sedimentation equilibrium* method is technically more difficult but is independent of shape and hydration of the particles, and directly yields average molecular weight data for the material under study. However, most viruses are unfavorably large for sedimentation equilibrium analysis.

3. *Electrophoresis (and Ion Exchange Chromatography)*

All viruses move characteristically in the electric field, owing to their ionizing groups. Their charge is obviously dependent on the pH of the solution. Thus below pH 4 all typical carboxyl groups are undissociated (—COOH, not $—COO^-$) and do not contribute to electrophoretic mobility; the same is true for typical amino groups above pH 11 ($—NH_2$, not $—NH_3^+$). However, with proteins and particularly with larger aggregates such as viral particles, electrophoretic mobility is determined almost exclusively by the polar groups situated at or near their surface, rather than by their total composition. This is illustrated by the fact that the empty capsid of TMV, the rod composed only of protein, shows the same electrophoretic mobility as the intact virion, even though the nucleic acid of the latter contributes 6390 additional negative charges to the particle. On the other hand, strains of TMV containing one more acid or basic group per protein subunit near the surface of the rod show detectable differences in electrophoretic mobility. Finally, the electrophoretic behavior of the free structural subunit is quite different from that of the capsid composed of the same protein. Not only is electrophoretic mobility affected by the spatial proximity of charged groups to the surface of the particle, but it is also determined by the density of charged groups on the surface, as contrasted to their number per unit of mass. Thus electrophoresis is often but not always a useful tool for the differentiation of viruses and virus strains and their components (259, 1, 14, 44, 213, 482, 342).

Notwithstanding the limitation of the theory and predictability of electrophoresis as applied to large molecular complexes, several electrophoretic techniques are available and have different advantages. The classical moving boundary methodology has been largely supplanted by zone electrophoresis. Filter paper moistened with a suitable buffer is frequently used, but starch or cellulose acetate gels have advantages. Polyacrylamide gels are now frequently preferred but are probably suitable only for small viruses (see Section IV,C).

Sucrose gradients have been used as supporting and stabilizing media in the zone electrophoresis of viruses (482). A capillary layer of buffer between two glass plates which are in contact with two lateral electrode vessels can serve for the electrophoretic separation of viruses by continuous free flow without supporting medium. Good separations of strains of small DNA viruses were obtained by both polyacrylamide gel and free flow electrophoresis (213, 44).

Methods utilizing ion exchange resins, celluloses, or gels rely on the same principle. Other column methods differentiate primarily on the basis of particle size by using more or less inert gels made up of beads of different porosity. Thus the principle of Sephadex fractionation has been extended to particles of virus dimensions by using agar or preferably agarose instead of dextran gels (3, 434). In all methods with gels, however, ion exchange plays a more or less strong contributing role.

4. *Serological Methods*

Immunological methods offer another tool of great importance in virus characterization. All viruses, whether plant, animal, or bacterial, are effective antigens when injected into rabbits or other suitable mammals. As in regard to the electrophoretic properties of viruses, the specificity of the antibodies formed is mainly a reflection of the surface character of the virus particle used in the immunization (347, 294, 483). In the case of the complex viruses containing envelopes, such as the myxoviruses (see Chapter VII), the antibodies react primarily with the proteins or lipoproteins of the envelope. Only by prior degradation of the envelope, which can be achieved by treatment with ether, are the antigenic activities of the capsid proteins revealed. In the case of the simple viruses the antibodies elicited by the intact virion are different from those elicited by the isolated structural subunits. As with electrophoresis, the antigenic specificity of the virion is affected by amino acid replacements or modifications near the virus surface, but not by similar alterations deeper within the particle.

The principal means of detecting antigen-antibody reactions are (a) precipitation and agglutination, including the formation of lines of precipitation appearing on agar plates between the wells containing antibodies and antigens, the two diffusing or electrophoresing toward one another; (b) neutralization of viral infectivity by the specific antibody; (c) complement fixation, removal of the normal serum component, complement, as a consequence of the interaction of antigen and antibody; (d) hemagglutination (see Chapter VII); (e) detection of the presence of viruses or viral components microscopically or electron microscopically through their reaction with a specific antibody. The antibody is made visible by coupling it with fluorescein, or electron-opaque by coupling it

---

FIG. 3-2. Indirect immunoferritin technique. Virus-infected cells fixed and reacted with specific antiviral antibody followed by ferritin-labeled antiglobulin. Ferritin molecules are separated from the viral particle by a space occupied by specific antibodies. (*a*) Poliovirus, extracellular, adsorbed to membranous debris. Note lack of ferritin on plasma membrane and membranous debris. ×130,000. (*b*) Sendai virus. Note ferritin on plasma membrane at site of altered membrane where virus will bud off. ×50,000. (*c*) Poliovirus, intracellular, among membranous vesicles. Both full and empty particles (arrows) ringed with ferritin. ×115,000. (*d*) SV40 virus. Virus emerging from disintegrating cell. ×63,000. (*e*) Human adenovirus type 12; virus in cytoplasmic debris. ×90,000. (*f*) Sindbis virus; virus adsorbing, 4 min after inoculation. ×77,000. (Courtesy of J. D. Levinthal.)

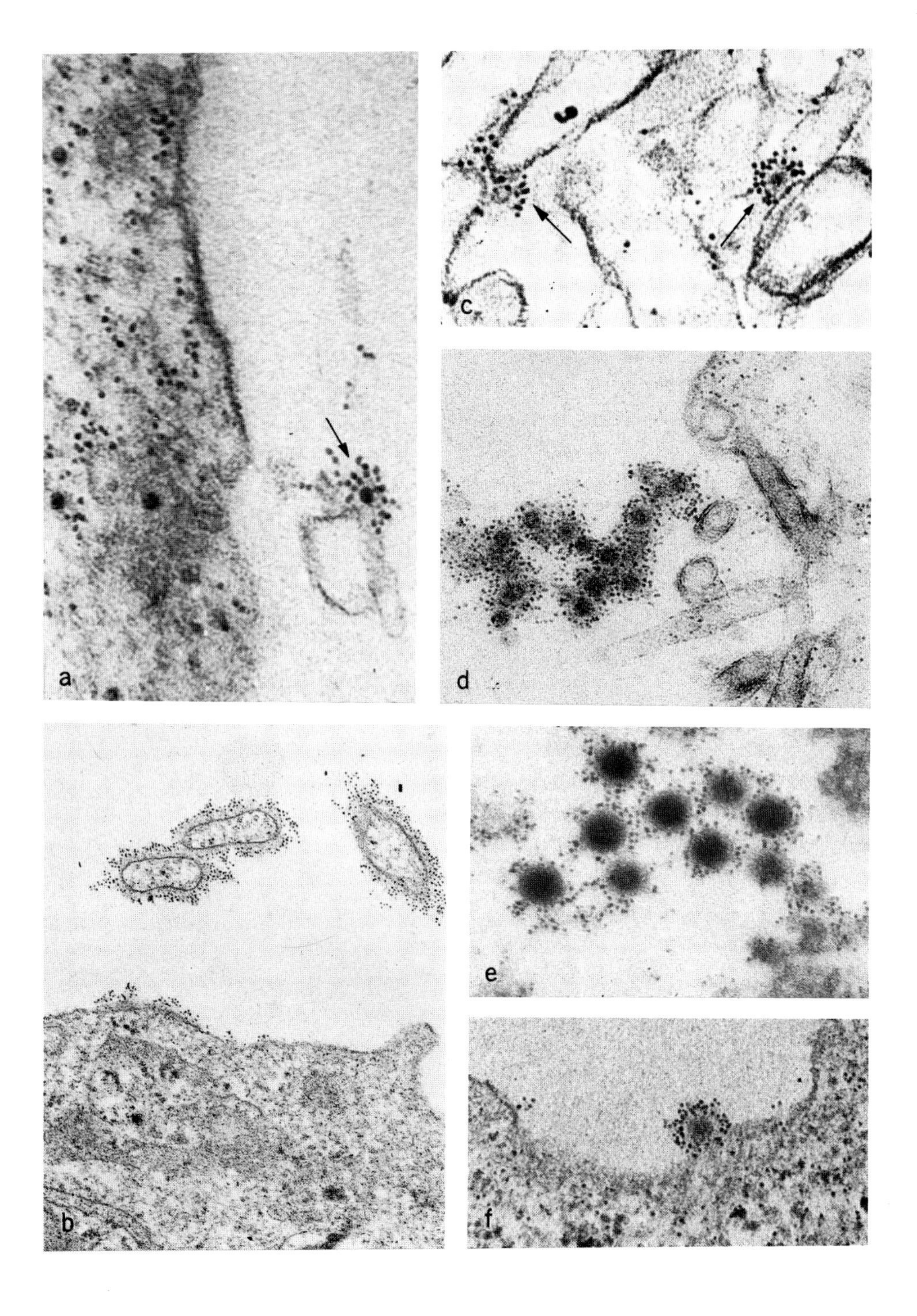
a
b
c
d
e
f

with ferritin, an iron-rich protein showing a typical pattern on electron micrographs (Fig. 3-2). Excellent reviews of the use of serological methods in virus research are available.

D. Tests for Purity of a Virus

With macromolecules or macromolecular complexes a strict demonstration of the identity of every molecule or particle is impossible. It is now known that individual variations exist, in terms of a very few residues, in the amino acid sequence of mammalian proteins such as hemoglobin; thus pure hemoglobin, unless isolated and crystallized from an individual cow's blood, is not strictly homogeneous. Yet it behaves as a homogeneous material in a long battery of tests, and the not quite perfect analytical stoichiometry can readily be attributed to the inherent error of the analytical methods.

In the same sense, the most highly purified virus preparations can only be regarded as approaching homogeneity in a statistical sense. It is nevertheless important to establish the degree of such homogeneity or heterogeneity as carefully as possible before using a virus for physicochemical, biochemical, or biological experiments. Most of the methods of purification represent at the same time means to ascertain the homogeneity of a virus. If a virus forms a single sharp band in a sucrose gradient, this indicates approximate homogeneity in terms of sedimentation properties. However, if the virus is purified only by repeated cycles of sucrose gradient centrifugation, faster or slower sedimenting material being removed each time, the uniformity of sedimentation has little discriminatory significance. In other words, only the use of a variety of methods relying on different principles in the course of purification and/or testing for homogeneity can supply significant evidence that we are dealing with a particular type of homogeneous particles. Such different criteria are sedimentation behavior, buoyant density, electrophoretic mobility, electron microscopic appearance, infectivity, stability (all previously discussed), and UV absorbancy.

The plotting of the *UV absorption spectrum* of a virus has become technically easy, since recording spectrophotometers have become available in most laboratories. Only 0.05–0.2 mg of virus is needed, and the material is not used up by this test. The details of the shape of the absorption curve reveal much information about the approximate composition of the virus in terms of protein and nucleic acid. Both the 260/280 mμ absorbancy ratio and the max/min ratio represent measures of the proportion of nucleic acid to protein in the sample, and a constancy in the details of the spectrum on subjection of the virus preparations to further purification and fractionation attempts is an additional indication that the virus is pure (see Section IV,C).

CHAPTER IV

METHODS FOR PREPARATION AND CHARACTERIZATION OF THE COMPONENTS OF VIRUSES

A. Methods of Degradation for Protein Isolation

The methods for the isolation of viral proteins would in principle be the same as those used for the isolation of viruses as well as of other typical proteins, were it not that a dissociating, disaggregating, and thus partly denaturing step must first release the protein from its multiple bonding to other protein molecules and to nucleic acid.

Although the denaturation of proteins is generally reversible, different proteins vary greatly in their tendency to reform the original conformation when the denaturant is being removed. Since the competing nonspecific aggregation reactions usually cause insolubility, thus shifting the equilibrium, many proteins are not readily renaturable in practice. Hence the dissociation of virus proteins should be performed with minimal concomitant denaturation.

Another aspect to consider is the masked state of the most reactive protein groups, generally the —SH groups, in native proteins and their susceptibility to autoxidation when denatured. Conditions for dissociation of viruses should be such as to minimize the danger of oxidative formation of disulfide linkages which would seriously interfere with the isolation of pure virus proteins of uniform physicochemical properties.

1. *Utilizing Neutral Media*

Labile viruses, such as the broad bean mottle, bromegrass, and cucumber mosaic viruses (BBMV, BMV, and CMV), which are degraded at neutrality by high ionic strength, would appear to be ideally suited for the purposes of protein isolation. However, the proteins of these viruses also have a particularly labile conformation, and thus they tend to become altered on storage even if autoxidation is prevented by the presence of dithiothreitol. The procedure for the isolation of these proteins is to dialyse the virus solution against 1 *M* $CaCl_2$ or 1 *M* NaCl buffered to pH 7 and containing 10^{-2} *M* dithiothreitol. The RNA then precipitates and can be removed by centrifugation (528).

Most viruses are dissociated at or near neutrality only if strong denaturants are added. Addition of urea or formamide to 8–10 *M* at pH 7–8 dissociates many viruses, and the protein can often be separated from the RNA by ammonium sulfate precipitation from such solution. TMV protein so isolated at 20% saturation with ammonium sulfate (SAS) is insoluble even if dithiothreitol is used to prevent autoxidation. Denatured TMV protein is successfully renatured if redissolved in 8 *M* urea and dialysed against 0.02 *M* phosphate of pH 6 (7). Renaturation can also be obtained after treatment with guanidine halides (4 *M*), which are stronger denaturants, by passage through urea. Detergents such as sodium dodecylsulfate (SDS) tend to be bound quite firmly to proteins—a major disadvantage in their use for protein isolation.

2. *Utilizing Acid Media*

Many stable viruses can be degraded by high concentrations of organic acids at 0°, the preferred agents being 67% acetic acid (117) or formic acid (307). These acids act both through protonization of all charged groups and through their hydrogen-bond displacing power. The acid medium does not favor any deleterious protein group reactions. Thus these media represent the best all-purpose methods available, although they may not split certain viruses or may denature the protein of others in a not readily reversible manner. The procedure is to add 2 volumes of undercooled glacial acetic acid to 1 volume of a rather concentrated virus solution (0.5–5%) at 0°, mix and swirl the resultant viscous solution or gel occasionally, and wait for a floccular precipitate to appear. If no precipitate becomes evident after 2 hr at 0°, the temperature can be raised gradually to 20°. The precipitate, largely nucleic acid, is centrifuged off and the water-clear supernatant containing the dissociated virus protein is dialyzed. TMV protein and many other virus proteins aggregate and may precipitate as their isoelectric point is approached during dialysis; then they can be isolated by centrifugation (8000 or 25,000 rpm, depending on their concentration and state of aggregation). Such precipitates, when resuspended in a little cold water, can usually be brought into solution by careful addition of 0.1 *N* NaOH to a final pH of 7.5–8. Naturally, the experiment is a success only if the product is soluble in water or dilute buffers of a near-neutral pH, and if it shows the typical UV absorption spectrum of a protein, with a max/min ratio (usually at about 280 and 248 mμ) of at least 2.0, since the presence of traces of nucleic acid strongly affects the minimum (Fig. 4-1). A final high speed centrifugation is often performed to assure the removal of any traces of undegraded virus or other particulate material.

3. *Utilizing Alkaline Media*

A classical method for viral protein isolation utilizes controlled alkaline pH such as dialysis against pH 10.5 buffers at 0–4° (390, 391). This pH, in the case of TMV protein, does not expose the —SH group sufficiently to render it autoxidizable. Any undegraded virus is separated by high-speed centrifugation, and the protein is precipitated with addition of ammonium sulfate to one-third saturation. The effect of alkali on a typical isometric virus is discussed also in the next Section.

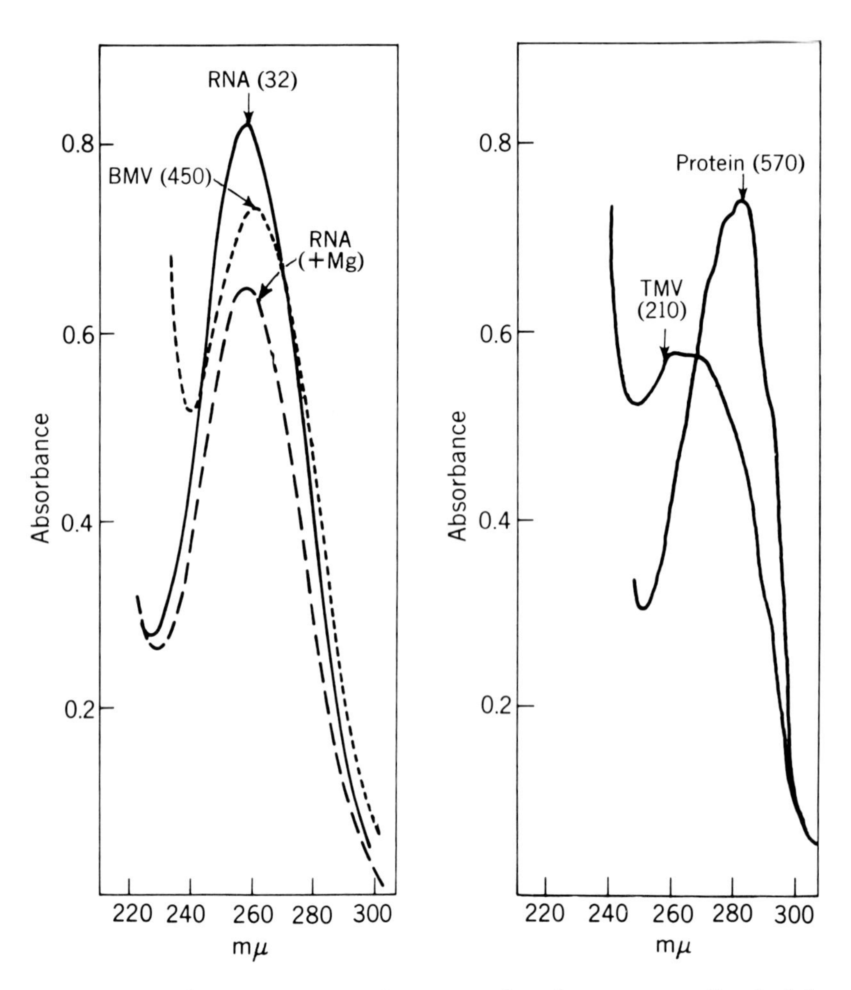

FIG. 4-1. Absorption spectra of viruses and viral components. On the left is an isometric virus containing 21% RNA (BMV), and viral RNA (TMV) in hyperchromed state in H_2O, and the same sample on addition of Mg Cl_2 to 10^{-3} *M*. On the right is TMV (whole virus, 5% RNA) and TMV protein. The numbers refer to the concentrations (in μg/ml) of the respective materials used. Note that the maximal absorbance per milligram is about 1:2:20 for TMV protein:TMV:TMV RNA.

4. *Utilizing Organic Mecurials*

Most virus proteins seem to contain cysteine residues, the —SH groups of which play some role in the conformation and interaction of the subunits. Thus treatment of different viruses with organic mercurials, reagents with a high affinity for —SH groups, has a variety of effects.

The one —SH group of the TMV protein is masked in the virus rod in such a manner that it is unreactive toward parachloromercuribenzoate (pCMB) but becomes substituted by the smaller methyl mercuric residue (116, 119). The rod structure is not affected by this reaction, and it is this mercury derivative that has played a key role in the elucidation of the helical structure of TMV (134) (see Section VIII,A).

In contrast, turnip yellow mosaic virus (TYMV), an isometric virus containing a protein with 4 —SH groups per subunit is dissociated by pCMB as well as by aliphatic mercurials (236, 238, 239). The same is true of poliomyelitis virus, another isometric virus, when it is treated with mercurials beyond a certain point (337, 64). On the other hand, the bushy stunt virus is not degraded by substitution with mercurials. On treatment with increasing concentrations of the particularly water-soluble mercurial, mersalyl, this virus appears to swell and finally to release its RNA in undegraded form (85).

In the case of TYMV the reaction with pCMB can be performed at pH 4.6 without apparent disruption of the particle but with loss of most of the infectivity of the virus. When this mercury-containing virus is brought to a neutral pH, however, it disintegrates. The loss of infectivity under these conditions is due to fragmentation of the RNA, but treatment of isolated TYMV RNA with pCMB does not cause its degradation (238). This is one of many indications that the protein of isometric viruses is closely integrated with the RNA. The older ideas of a protein shell around an RNA core have been displaced by the present concept of RNA loops passing through each of the 32 capsomeres making up the virus particle. This concept is based primarily on recent X-ray diffraction studies with TYMV, but it is also supported by studies of the degradation of that virus, and it may be generally applicable to the considerable number of viruses resembling TYMV in their architecture (251) (see Section VIII,B).

The effect of alkali on TYMV is also quite complex, but it finds partial explanation in present structural concepts concerning these viruses. Apparently, alkali at low ionic strength largely degrades the protein shell. However, above 0.5 *M* salt, the protein shells remain intact, while losing their nucleic acid and thus becoming artificial top component particles (ATC), to differentiate them from the naturally occurring RNA-free top component, NTC (237). In contrast to the virus, isolated ATC or NTC particles are resistant to alkali, even at low ionic strengths, a surprising fact that requires further investigation. A study of the RNA fragments obtained by alkaline degradation indicates a predominant component of 4.6 S and an approximate molecular weight of 57,000 (approximately 1/32 of that of intact TYMV RNA). The various data of this type are interpreted to indicate that in these viruses the RNA approaches the surface and is somewhat susceptible to alkali (and nucleases) at 32 sites associated with the 32 capsomeres (see Fig. VIII,B 8-7). Substitution of the —SH groups by mercurials further exposes the RNA. Once the RNA has become fragmented, it no longer stabilizes but rather destabilizes the particle, as does the mercury substitution. Comparable observations have been made with the poliomyelitis virus, and the results obtained with the bushy stunt virus are also in accord with such a mechanism (236).

B. Methods of Isolation of Viral Nucleic Acids

Some of the methods which degrade viruses are particularly suited for the preparation of native proteins, and others for that of intact nucleic acids. The most useful methods for protein isolation proceed in the cold but utilize low or high pH's, both of which endanger the integrity of nucleic acids. In contrast, the two methods which, at times in combination, have proved most generally useful for the isolation of infective nucleic acids, detergents and phenol, degrade viruses at neutrality, although at times requiring elevated temperatures. Heating of viruses in salt solution also has been used to produce infective RNA.

1. *Detergents*

Most viruses are degraded by treating them with 1% sodium dodecyl sulfate (SDS), although either slightly acid or alkaline pH or heating is usually necessary. TMV is split in a few minutes at 55° at pH 8.8 (128). Addition of a half volume of saturated ammonium sulfate precipitates the denatured protein, and the RNA separates from the supernatant on cooling or, more rapidly, on addition of more ammonium sulfate. The RNA precipitate is redissolved in a little water, and reprecipitated with 2 volumes of ethanol. It is then subjected to high-speed centrifugation if the removal of residual traces of undegraded virus is to be assured. RNA obtained in this manner is as free of protein as most RNA or DNA preparations, and it was the first RNA isolated from a virus in biologically active form (see Section VIII,H).

2. *Phenol*

The most popular and most generally applicable method of preparing RNA is the phenol method (147). The principal feature of this method is the homogenization of aqueous virus solutions, excess phenol being used as a dissociating and denaturing agent. Phenol also acts as a solvent for the released protein, while the nucleic acid remains in the aqueous phase. Phenol treatment can be performed either without salts except 0.02 *M* EDTA (173) (and then yields hyperchromed RNA) or in the presence of 0.1 *M* phosphate or tris buffers (pH 7) (then the product is hypochromed RNA). Usually the virus solution is stirred vigorously or homogenized with an equal volume of redistilled phenol saturated with water or buffer. The mixture is then centrifuged, and the aqueous solution pipetted off the phenol and the gelatinous interphase. The aqueous phase is treated a second time with phenol in the same manner. The phenol and interphase fraction can be backwashed with water or buffer if quantitative recovery is important. The (pooled) aqueous phases are treated with 2–3 volumes of ethanol and, in the absence of buffers, pH 6–7 acetate to approximately 0.05 *M* to flocculate the RNA. Precipitations (3–4) from water with ethanol, with thorough draining of the supernatants from the well-packed RNA, remove the phenol quantitatively and render the commonly used ether extractions unnecessary.

Phenol treatment is usually performed at room temperature, although 0° has been advocated at times (147), and elevated temperatures are necessary to extract the RNA from some sources, such as the virions of eastern equine encephalitis virus (503). For the isolation of the RNA from several plant viruses, as well as of the DNA from many viruses, it is advantageous to combine phenol extraction with a detergent treatment (0.1% SDS, 0.01% EDTA, 37°, 15 min). Many DNA viruses require pretreatment with a protease (e.g., pronase) to obtain good yields of DNA by either phenol or detergent treatment or both.

RNA is generally quite susceptible to nucleases, and in the case of viral RNAs of the size of TMV RNA, in which one split per 6400 nucleotides represents an inactivating event, the presence of traces of nucleases or phosphodiesterases can spell the difference between success and failure in experiments with naked viral RNA. Various means have been advocated to combat this danger. Pretreatment of TMV with chelating agents, e.g., by incubation with citrate or EDTA, frees the virus largely from nucleases that usually adhere to it. More generally applicable is the use of *nuclease inhibitors* during the isolation of the RNA. Bentonite, a polyacidic clay, appears to be at least as effective as other polyacidic minerals or polymers advocated later (129, 419). Washed and graded bentonite may be added in suspension (at about 10–30% of the weight of the virus) during the phenol procedure. The finer particles remain suspended in the aqueous phase but are completely removed by high speed centrifugation of the RNA after partial or complete removal of the ethanol.

C. Methods of Characterizing Proteins and Nucleic Acids and Tests for Their Homogeneity

Most of the techniques used for the characterization of viral proteins and nucleic acids are the same as those used with whole viruses, and they have been discussed in Chapter III. As for whole viruses, the *UV absorption spectra* represent a most useful tool for the characterization of viral proteins and nucleic acids (Fig. 4-1).

The minimum in protein absorption at about 248 mμ is strongly affected by the presence of nucleic acid, and the minimum in the nucleic acid spectrum near 228 mμ is an even more sensitive indicator of the presence of proteins all of which absorb highly in that region (128). The details of the absorption curve of proteins at neutrality, and the effect on their absorbance of adding alkali to 0.1 *M*, can be used for estimation and quantitation of their tyrosine and tryptophan content (118). With proteins and nucleic acids the absorbance also gives clear indications regarding their conformation. Thus denaturation of proteins usually causes detectable changes in the spectrum (mainly a slight shift of the maximum to lower wavelengths). Nucleic acids can be observed in both the hyperchromed and the hypochromed states; they show temperature- and ionic strength-dependent differences of 20–40%. Finally, most methods of base analysis utilize the characteristic absorption spectra of the components of nucleic acid for both purposes of identification and quantitation (see Section VI,A).

Density gradient centrifugation in CsCl and similar salts of high density is particularly useful in studies on nucleic acids because it permits the differentiation between "heavy" and "light" nucleic acid fractions or strands, the "heavies" being produced by growing the virus or organism in the presence of bromouracil (instead of thymine) (153) or the heavy isotopes of H, N, and O (^{2}H, ^{15}N, ^{18}O). The buoyant density of a double stranded DNA in CsCl is also a function of its base composition, and thus analytical data (% G + C) can be derived from equilibrium density gradient centrifugation (446) (Fig. 4-2) (see Section VI,A). Sucrose gradients prepared in D_2O can also be used for this purpose. The high resolving power of equilibrium density gradient centrifugation, which makes it possible to separate the two strands of DNA as well as the two halves of fragmented DNA of certain viruses (see Section VI,D) (185, 208), have made this a most powerful tool of current research.

Another technique that is presently much favored is *polyacrylamide gel electrophoresis*. By addition of varying amounts of

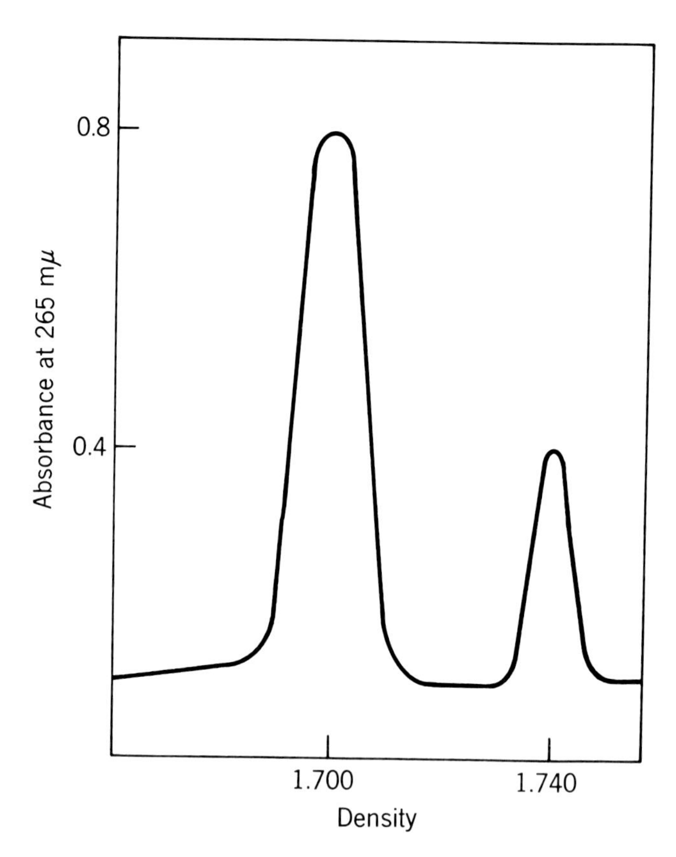

Fig. 4-2. Cesium chloride gradient separation of a single stranded and a double stranded DNA. The double stranded polymer sediments in the lower density region.

crosslinking monomers (e.g., methylene bisacrylamide, ethylene diacrylate) before polymerization of the acrylamide in glass tubes, varying extents of crosslinking can be achieved in these gels (124, 373). Their resistance to denaturing solvents such as 8 *M* urea or 1% sodium dodecyl sulfate (SDS) represents another important advantage. By applying a potential gradient, proteins, peptides, nucleic acids, and viruses can be made to migrate characteristic distances into such gels, their migration depending largely on their molecular or particle weight and on the extent of crosslinking of the gel. The materials are subsequently detected as bands either by staining or local precipitation methods or by radioautography of the gels if radioactive substances are being separated (see Section XI,B). The power of resolution is sufficient to reveal and identify approximately 37 species of ribosomal proteins (470). The doubts about whether the multiple bands represent true differences or are due to artifacts have now been largely answered and abandoned. Nevertheless, it has also been recognized that different bands do not necessarily represent completely different materials but may actually be as closely related as the same protein with one amide ($—CO—NH_2 \rightarrow COO^-$) or one acetyl group ($—NH_3^+ \rightarrow —NH—COCH_3$) more or less (123). Polyacrylamide gel electrophoresis has yielded most of our current data on the multiple nature of the proteins of certain scarce viruses. Its application to nucleic acids and viruses has yielded similarly definitive results.

The principal factor determining mobility during polyacrylamide gel electrophoresis, particularly as usually performed with proteins dispersed and analyzed in 0.1% SDS, is not the characteristic charge of the macromolecule. Under these conditions the basic groups of proteins form SDS complexes, and proteins, like nucleic acids, behave largely like polyanions. Separations are then due mainly to differences in molecular weight, the smaller components moving ahead of the larger. By standardizing such gels with proteins or nucleic acids of known molecular weight, that of unknowns can be determined with reasonable accuracy (401). It must be noted particularly that the common procedure of performing these runs in the presence of strong denaturants and reducing agents yields data on the molecular weights of structural subunits or even of peptide chains if a given protein consists of two or more chains held together by disulfide bonds. However, it appears probable that in urea, rather than SDS, and at a pH near the isoelectric point true electrophoretic mobility plays a contributing role in determining the relative position of different proteins or peptides. The application of polyacrylamide gel to the fractionation of nucleic acids and their fragmentation products appears particularly promising (35, 82, 333).

CHAPTER V

THE PROTEINS OF VIRUSES

A. General Methods

In general terms, virus coats or capsids are made up of relatively small proteins of a regular geometry which enables them to aggregate to a stable shell of definite dimensions. Ionic interactions, hydrogen bonds, and hydrophobic interactions, but no primary bonds, are involved. The structural requirements for virus proteins which enable them to bind identical, or at times also not identical, molecules in a highly specific manner are no less complex than that of enzymes or specific binding proteins such as avidin and antibodies (emphore proteins) (330), which bind substrates or other small or large molecules with high affinity and specificity. Such binding capabilities are a consequence of specific protein conformations, and protein conformation is dictated by amino acid sequence. Thus the analysis of the amino acid sequences of virus proteins is of considerable interest, and such studies have been performed in a number of laboratories and on a variety of viruses. Apart from the interest of protein structure in terms of function, virus protein sequences are also of interest as chemically defined phenotypic expressions of the corresponding viral genomes.

TABLE 5-1

THE SIDE CHAINS OF THE AMINO ACIDS WHICH MAKE UP TYPICAL PROTEINS

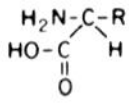

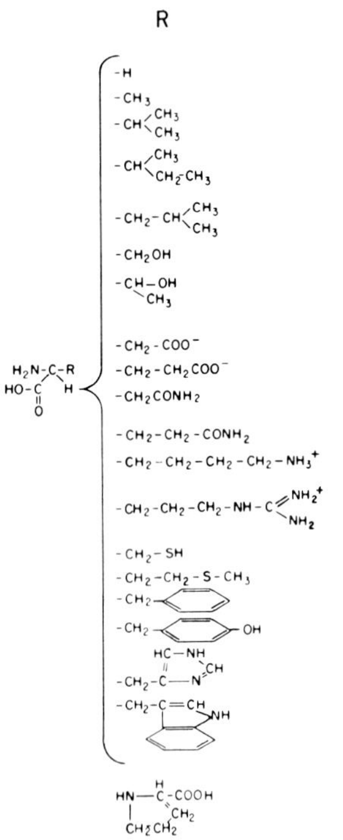

R	Name	Abbreviation	Codons
$-H$	Glycine	(Gly)	GG(Pu,Py)
$-CH_3$	Alanine	(Ala)	GC(Pu,Py)
$-CH(CH_3)_2$	Valine	(Val)	GU(Pu,Py)
$-CH(CH_3)-CH_2-CH_3$	Isoleucine	(Ile)	AU(Py,A?)
$-CH_2-CH(CH_3)_2$	Leucine	(Leu)	UUPu, CU(Pu,Py)
$-CH_2OH$	Serine	(Ser)	UC(Pu,Py), AGPy
$-CH(OH)-CH_3$	Threonine	(Thr)	AC(Pu,Py)
$-CH_2-COO^-$	Aspartic acid	(Asp)	GAPy
$-CH_2-CH_2COO^-$	Glutamic acid	(Glu)	GAPu
$-CH_2CONH_2$	Asparagine	(Asn)	AAPy
$-CH_2-CH_2-CONH_2$	Glutamine	(Gln)	CAPu
$-CH_2-CH_2-CH_2-CH_2-NH_3^+$	Lysine	(Lys)	AA Pu
$-CH_2-CH_2-CH_2-NH-C(=NH_2^+)NH_2$	Arginine	(Arg)	CG(Pu,Py),AGPu
$-CH_2-SH$	Cysteine	(Cys)	UGPy
$-CH_2-CH_2-S-CH_3$	Methionine	(Met)	AUG
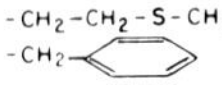	Phenylalanine	(Phe)	UUPy
	Tyrosine	(Tyr)	UAPy
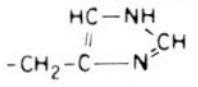	Histidine	(His)	CAPy
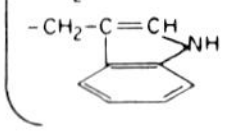	Tryptophan	(Trp)	UGG
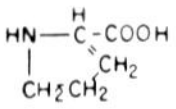	Proline	(Pro)	CC(Pu,Py)

Peptide chain formation

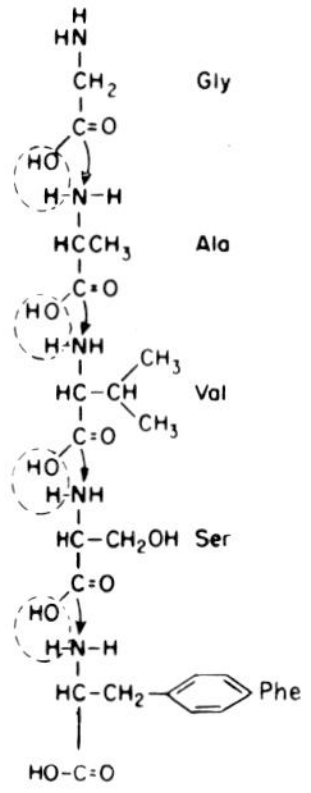

Table 5.1 may refresh the reader's memory of the chemical nature of the R groups, the side chains of the amino acids which make up the typical proteins, including viruses. Since the α-amino and α-carboxyl groups are condensed to peptide linkages as illustrated in Table 5.1, the nature of the R groups determines the character of the proteins. The customary three-letter abbreviations are also listed in this Table, as well as the RNA codons corresponding to each amino acid (see Section IX,B).

The methods of amino acid sequence analysis used with virus proteins are the same as those used with other proteins and are described here only briefly. The reader is referred to other sources for more detail (392, 124).

1. *Amino Acid Analysis*

Analyses for amino acid composition after acid hydrolysis are very frequently employed in the course of sequence analysis, and the use of an automatic analyzer is now the method of choice. However, frequently the necessary precautions in the preparation of the hydrolysates are not observed, and this can lead to errors. With careful sealing of evacuated tubes containing 1–5 mg of protein and redistilled 6 *N* HCl, and well-controlled heating at 108°, reproducible results are usually obtained. However, various proteins or peptides differ in the rate with which serine, threonine, and sometimes tyrosine decompose during hydrolysis. The rate of release of isoleucine, leucine, and valine also differs. Only hydrolyses performed for several time periods and extrapolation of amino acid decompositions to zero-time hydrolysis can give approximately correct values for the labile amino acids, and the maximal values obtained for the hydrophobic amino acids, usually after 72 hr, approach their actual presence.

2. *End Group Analysis*

Analyses for the number and nature of end groups in viral proteins and their degradation products are frequently of considerable value in their characterization. The preferred methods are the fluorodinitrobenzene (FDNB), dansyl, and phenylisothiocyanate (PTC) methods for the amino end, and hydrazinolysis for the carboxyl end (Figs. 5-1, 5-2). Enzymes that degrade peptides and proteins from the amino end (leucine aminopeptidase) or the carboxyl end (carboxypeptidase A and occasionally B), can also be used to determine terminal sequences.

Since most viral proteins lack terminal amino groups because these are acylated, analyses for such acyl groups (usually acetyl) must frequently be performed. The initiation of *E. coli* proteins, including those of *in vitro* biosynthesized phages, with *N*-formylmethionine (see Section IX,B), also calls for acyl analyses. The most direct but not necessarily the most reliable method for this purpose is gas-chromatographic analysis to detect acetic acid or formic acid

$Prot{-}NH_2 + FC_6H_3(NO_2)_2 \longrightarrow Prot{-}NH{-}C_6H_3(NO_2)_2 + HF$

(1,2,4-fluorodinitrobenzene, FDNB)

(Yellow DNP dinitrophenyl, derivative),

+ 6 N HCl (4-12 hr, 108°)

$HOOC{-}CHR{-}NH{-}C_6H_3(NO_2)_2$ + amino acids and peptides

(Yellow DNP amino acid from amino end of chain, ether-soluble from acid if R is not basic)

$Prot{-}NH_2$ + dansyl (diaminomethylnapthalylsulfonyl) chloride

Strongly fluorescent amino acid derivatives
(similar to FDNB, but more sensitive, and derivatives more stable)

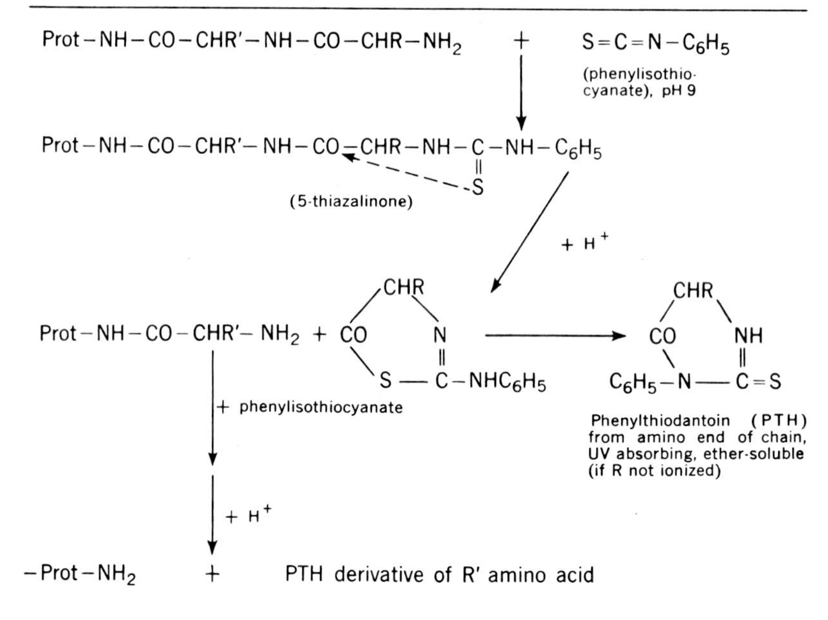

FIG. 5-1. N-Terminal analysis of proteins. Reaction of protein amino groups with fluorodinitrobenzene (or dansyl chloride) gives yellow (or fluorescent) nonbasic derivatives only of the amino terminal residue. Phenylisothiocyanate treatment, followed by acid, allows stepwise degradation of amino acids from the N terminus, yielding phenylthiohydantoins of the released amino acids.

$$H_2N-CHR-CO-NH-CHR^1-CO-NH-CHR^2-CO-NH-CHR^3-COOH$$

↓ + anhydrous hydrazine (NH_2-NH_2) 80°, 12 hr

$H_2N-CHR-CONHNH_2$

$H_2N-CHR^1-CONHNH_2$ + NH_2-CHR^3-COOH

$H_2N-CHR^2-CONHNH_2$ Only one free amino acid from carboxyl end

FIG. 5-2. Hydrazinolysis. Degradation of proteins with hydrazine yields only the C-terminal amino acid in free form.

in the hydrolysate (308). The protein chemist's approach is to isolate and purify the only peptide lacking a basic group after peptic, chymotryptic, or thermolytic (not tryptic; see below) digestion of the acylated protein (Fig. 5-3). The terminal peptide is the only one that does not "stick" on a Dowex 50 column at acid pH (see later). Its composition, including the nature of its acyl group, can

$$CH_3CO-NH-CHR-CO-NH-CHR^1-CO-NH-CHR^2-CO-NH-CHR^3-CO-\ldots$$

+ enzyme that splits preferentially after R^2 amino acid

(e.g., $R^2 = -CH_2C_6H_5$, enzyme = chymotrypsin)

↓

$CH_3CO-NH-CHR-CO-NH-CHR^1-CO-NH-CHR^2-COOH$ (one peptide lacking free terminal amino groups, and many peptides of various lengths carrying free terminal amino groups)

FIG. 5-3. Detection of terminal acylation. Enzymatic degradation of a terminally acylated protein yields a nonbasic peptide only from that acylated end.

be determined by a combination of amino acid analysis, hydrazinolysis, and the FDNB method.

3. *Multiple Chains*

The presence of more than one chain is tested for by studying the molecular weight of the protein under denaturing conditions, i.e., in 1% sodium dodecyl sulfate (SDS) directly and after reduction of any disulfide bonds by mercaptoethanol or dithiothreitol, preferably followed by alkylation of the —SH groups by iodoacetate, iodoacetamide, or ethylene imine (Fig. 5-4). Polyacrylamide gel electrophoresis (see Section IV,C) of such preparations represents a convenient tool, particularly in SDS-containing buffers, for obtaining approximate molecular weight data and detecting subunits or fragments, if the protein of a virus is composed of multiple chains,

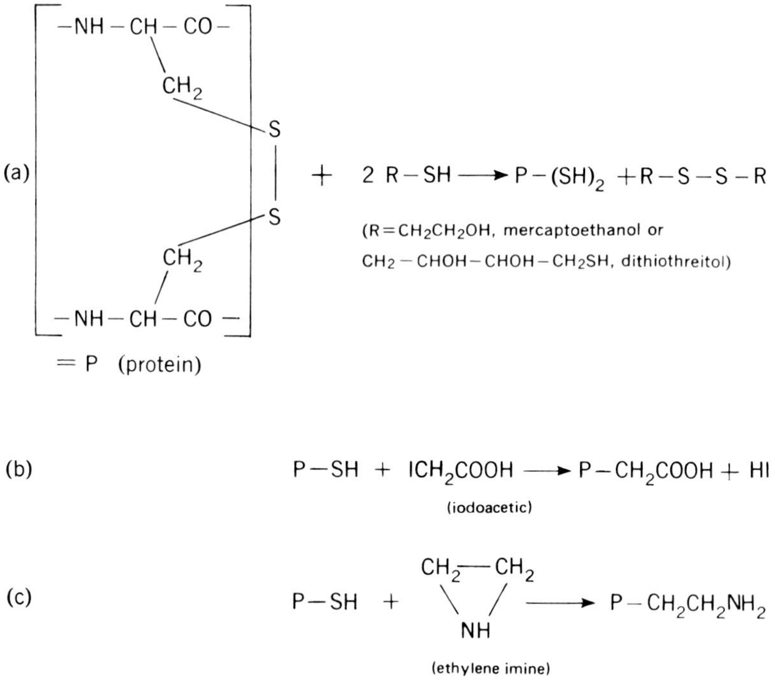

(a) $\left[\begin{array}{c} -NH-CH-CO- \\ | \\ CH_2-S \\ | \\ CH_2-S \\ | \\ -NH-CH-CO- \end{array}\right] + 2\ R-SH \longrightarrow P-(SH)_2 + R-S-S-R$

= P (protein)

($R = CH_2CH_2OH$, mercaptoethanol or $CH_2-CHOH-CHOH-CH_2SH$, dithiothreitol)

(b) $P-SH + ICH_2COOH \longrightarrow P-CH_2COOH + HI$

(iodoacetic)

(c) $P-SH + \underset{NH}{CH_2-CH_2} \longrightarrow P-CH_2CH_2NH_2$

(ethylene imine)

FIG. 5-4. Reduction-alkylation. Reduction (*a*) of protein disulfide groups is achieved by addition of an excess of a mercaptan. The (resulting) protein —SH groups can be alkylated with a variety of compounds (*b*, *c*).

be they free or held together by secondary forces or by disulfide bonds (cystine residues) (see Section V,B,3). Obviously the end group analyses and the peptide chain weight should give concordant results.

4. *Selective Splitting of Peptide Bonds*

If it is assumed that the protein consists of a single chain (as it usually does), or that previous fractionations have resulted in single chain preparations, the next step is selective degradation of the polypeptide. Digestion with trypsin is most useful for this purpose since it breaks only peptide bonds at the carboxyl side (at the right) of most arginine and lysine residues, as well as at cysteine or cystine residues if they have been transformed to the S-ethylamino derivatives (with an R group resembling that of lysine) by ethylene imine treatment (see Figs. 5-4, 5-5) with or without prior reduction. Alternative enzymatic means of degrading the polypeptide are: chymotrypsin, which splits preferentially, though not selectively, at the right of tyrosine, tryptophan, phenylalanine, and leucine residues; and the bacterial enzyme, thermolysine, which splits

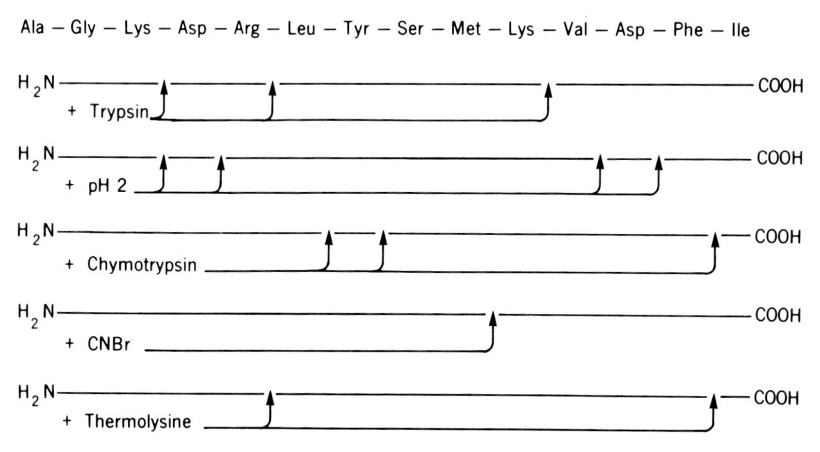

FIG. 5-5. Selective splitting of a peptide. The preferred points of attack of various more or less specific peptide bond-breaking agents on a peptide are indicated. From amino acid and some end group analysis data on peptide fragments obtained by two or three of these methods the entire sequence can be deduced.

preferentially at the left of leucine and isoleucine residues. A useful nonenzymatic method is degradation in acid solution (pH 2–3) which on prolonged heating breaks on both sides of aspartic acid, and particularly degradation by cyanogen bromide (CNBr) which in acid at room temperature splits the peptide bonds on the right of methionine.

5. *Fractionation of Peptides*

All degradative methods yield mixtures of peptides which are shortest, on the average, after chymotrypsin, and longest after CNBr treatment (because the methionine content of most proteins is quite low compared to that of the sum of the large hydrophobic amino acids). The fractionation of such mixtures is best achieved on ion exchange columns. Dowex 1, and related basic resins, eluted with pyridine acetate buffers of gradually decreasing pH (8–3), let the most basic peptides emerge first, and the most acidic ones last, while Dowex 50 type resins (with buffers from pH 3–8) give the opposite order of elution (Fig. 5-6). In both cases large peptides and peptides rich in aromatic residues are retarded. The peptides are detected by the ninhydrin test (specific for amino groups) performed on aliquots, preferably after alkaline hydrolysis to change them to their constituent amino acids, or by the Folin biuret test (which measures peptide bonds and tyrosine). After amino acid analysis it is often desirable to do end group analyses on such peptides, which in the case of di- and tripeptides establish the sequence. Frequently the Edman method of stepwise degradation with phenylisothiocyanate (see Fig. 5-1) is very useful in determining the sequence from the amino end, a method which may also be used with the undegraded viral proteins, if an amino terminal residue can be detected. An automatic "sequenator" utilizing this reaction is becoming commercially available. Figure 5-5 illustrates the specific breakage of a pentadecapeptide the sequence of which is established by these degradations and end group analyses alone.

Often it is advantageous to further degrade the larger peptides individually by one of the other methods listed above and to determine the amino acid sequence of the fragments. If the sequences of all of one group of peptides, e.g., those resulting from tryptic digestion, have been determined, then the composition of

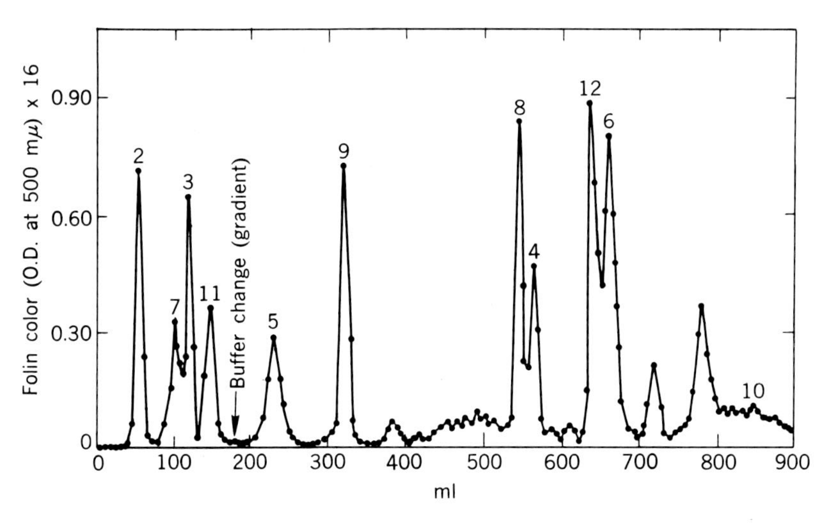

FIG. 5-6. Separation of tryptic peptides on Dowex 1. TMV protein digest fractionated into the component peptides. The large N-terminal peptide (#1, residue 1-41) has been removed as a precipitate at pH 4.7. The numbers of peptides are in order from the N terminus to the C terminus. Peptides 2, 7, 3, 11, 5, and 9 have 5, 2, 15, 7, and 3 amino acid residues and +1, +1, +2, +1, and +1 net charges, respectively. The others are neutral or negatively charged. The peptides are not necessarily pure at this stage.

peptides obtained by another method usually enables the investigator to establish a working hypothesis for the sequential arrangement of the first group of peptides. This arrangement can then be confirmed by sequence analysis of some of the peptides obtained by the secondary method of splitting. Thus, for example, the study of many arginine- and lysine-containing peptides in a chymotrypsin digest establishes overlap sequences to permit the sequential arrangement of tryptic peptides (see Fig. 5-5).

B. Specific Proteins

1. *Coat Protein of TMV*

The first viral protein of which the amino acid sequence was determined was the coat protein of TMV, and the minor differences and the few errors in the structures, as simultaneously elucidated by two laboratories, have now been corrected (Fig. 5-7) (9, 10, 145,

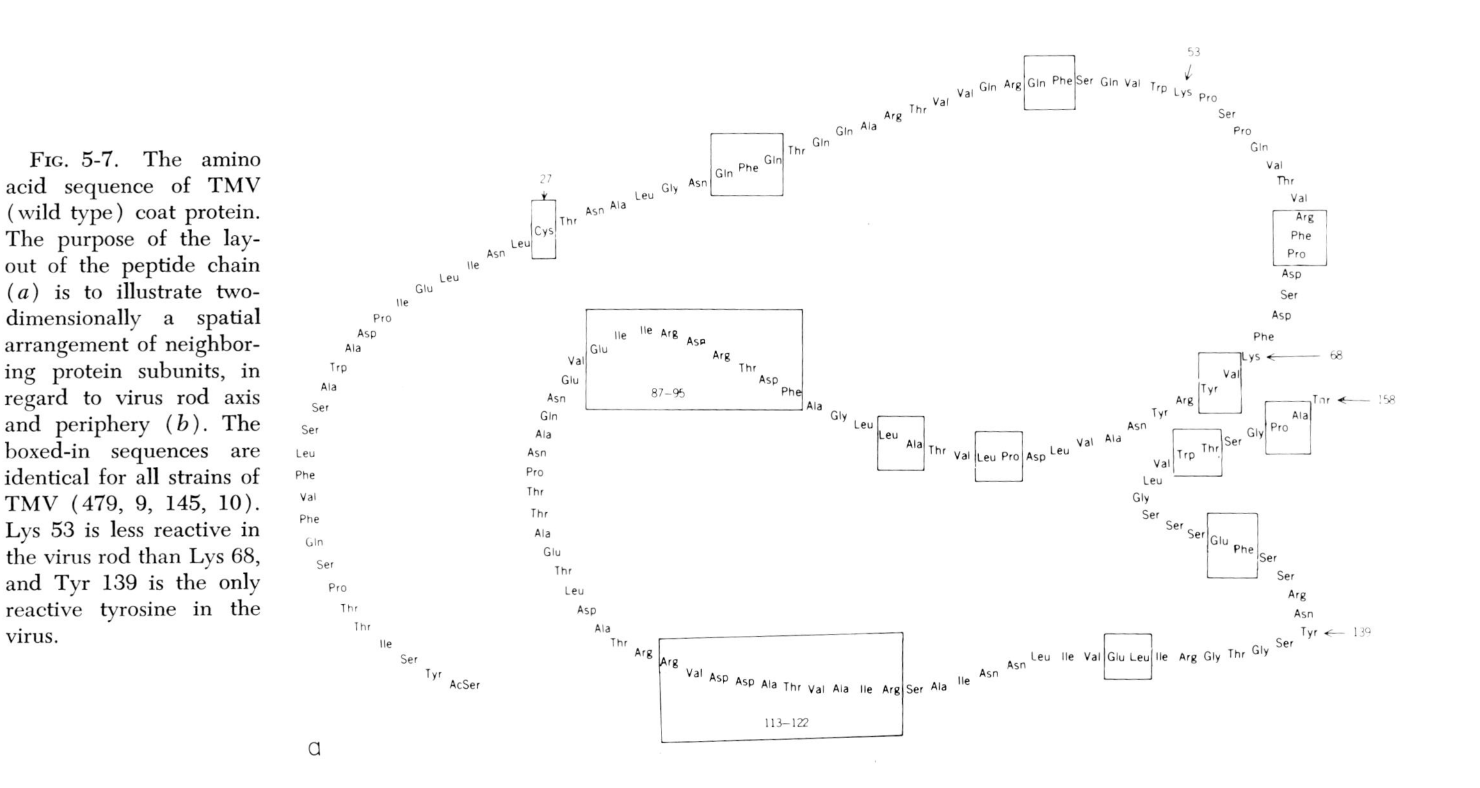

FIG. 5-7. The amino acid sequence of TMV (wild type) coat protein. The purpose of the lay-out of the peptide chain (*a*) is to illustrate two-dimensionally a spatial arrangement of neighboring protein subunits, in regard to virus rod axis and periphery (*b*). The boxed-in sequences are identical for all strains of TMV (479, 9, 145, 10). Lys 53 is less reactive in the virus rod than Lys 68, and Tyr 139 is the only reactive tyrosine in the virus.

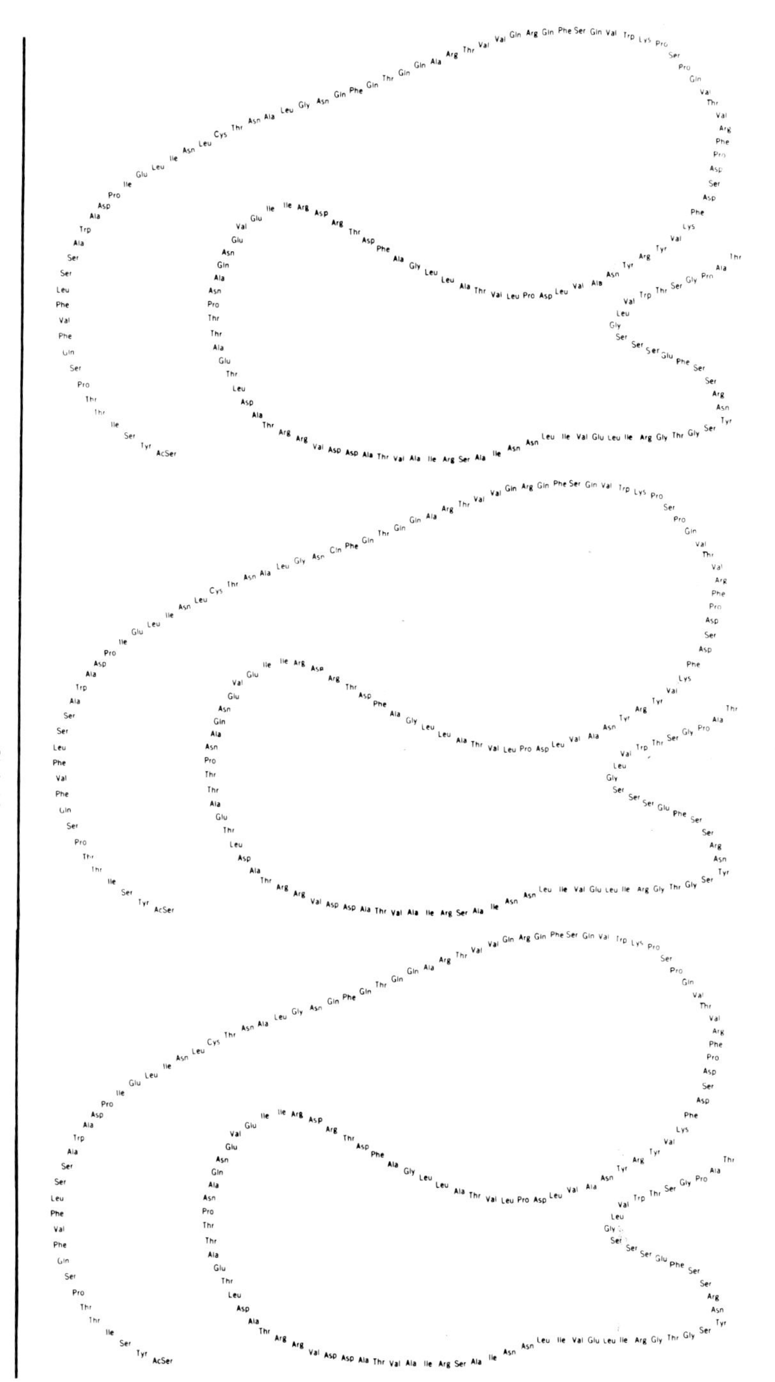
rod axis
rod surface
b
AcSer Tyr Ser Ile Thr Thr Pro Ser Gln Phe Val Phe Leu Ser Ser Ala Trp Ala Asp Pro Ile Glu Leu Ile Asn Leu Cys Thr Asn Ala Leu Gly Asn Gln Phe Gln Thr Gln Gln Ala Arg Thr Val Val Gln Arg Gln Phe Ser Gln Val Trp Lys Pro Ser Pro Gln Val Thr Val Arg Phe Pro Asp Ser Asp Phe Lys Val Tyr Arg Tyr Asn Ala Val Leu Asp Pro Leu Val Thr Ala Leu Leu Gly Ala Phe Asp Thr Arg Asp Arg Ile Ile Glu Val Glu Asn Gln Ala Asn Pro Thr Thr Ala Glu Thr Leu Asp Ala Thr Arg Arg Val Asp Asp Ala Thr Val Ala Ile Arg Ser Ala Ile Asn Asn Leu Ile Val Glu Leu Ile Arg Gly Thr Gly Ser Tyr Asn Arg Ser Ser Phe Glu Ser Ser Ser Gly Leu Val Trp Thr Ser Gly Pro Ala Thr

474). TMV protein is a single chain of 158 amino acids containing one cysteine —SH group (residue 27), but not cystine, nor methionine or histidine. The N-terminal serine is acetylated (317). End group acetylation has since been found to be common among plant virus proteins. This has represented a major stumbling block in the chemical study of these proteins, since acetylated proteins were not known to exist in 1952 when N-terminal analyses were initiated with TMV protein. The carboxy-terminal threonine also behaves unusually in that it is readily released by carboxypeptidase, even from the intact virus, while no further amino acids are split off by longer enzyme action (167). It is now clear that the location of proline in the third position from the end prevents enzymatic release of the alanine which becomes terminal on removal of the threonine. That proline plays such a protective role against exopeptidase action is illustrated by the properties of certain mutants of TMV in which the proline in position 156 is replaced by leucine: in contrast to wild type TMV, these mutants are susceptible to extensive degradation by carboxypeptidase (473). It appears that both the acetylation of the N terminus and the location of proline near the C terminus represent evolutionary advantages in giving the protein of wild type TMV additional stability.

A generally important factor in rendering proteins more stable and resistant to a variety of agents is their three-dimensional structure or conformation (tertiary structure). Proteins are usually more enzyme-resistant in the native state than after denaturation; i.e., after their specific conformation has been abolished. The multiple bonds formed during the aggregation of protein subunits to virus capsids (quaternary structure) contributes further to the masking of enzyme-susceptible sites and to stabilization against heat and other agents. This is well illustrated with TMV because the protein of the TMV particle is almost completely resistant to many typical proteases, which attack the disaggregated native protein (predominantly trimers of the structural subunit; see Section VIII,A) slowly and the denatured protein rapidly. Also, the protein in intact TMV is resistant to temperatures up to 60–70°, whereas the disaggregated protein becomes denatured at 35–40° and thus becomes susceptible to proteases at such temperatures.

Another striking difference in the properties of free TMV protein, as contrasted to the protein in the virus particle, is shown by its antigenic properties. In the virus the C-terminal segment and residues between 60 and 70 are the main determinants, and both are believed to lie near the surface of the rod (8, 11, 398). In contrast, the antibody to the isolated protein shows a particular binding affinity to residues 108–112 (531). This has led to a systematic study of the role of each residue in this binding using synthetic peptide analogs of this part of the TMV protein molecule (532).

From the sum total of the chemical, immunological, electrophoretic, and enzymological data now available (122) concerning the reactivity and thus the presumed surface proximity of different protein groups, the approximate spatial arrangement of certain segments of the TMV peptide chain can be deduced, as schematically represented in two-dimensional manner in Fig. 5-7. The definitive method of protein conformational analysis, small angle X-ray diffraction, has until recently been applicable only to proteins forming large three-dimensional crystals. However, recent advances in that methodology are expected to lead to a picture of the conformation of the TMV protein in the virus rod in the not too distant future.

Several plant viruses detected in the field or in the laboratory were found to be related to TMV and are regarded as strains of this virus. They all show rods of indistinguishable appearance on electron micrographs, and they show the same or very similar X-ray diffraction patterns as TMV (see Section VIII,A). They react to a different extent with anti-TMV serum. Amino acid analysis reveals that the strains cross-reacting most completely with antibody to the wild type differ in only 1 or 2 amino acid replacements. On the other hand, several groups of strains showing 8 or 17 net exchanges have been detected. Table 5.2 illustrates the typical composition of prototypes of the six classes now recognized (A–F), (122). Within several of these classes, variants showing exchanges of 1-3 residues have been detected. It will be noted that the classes show characteristically progressive features such as 4, 5, 6, and 7 tyrosines; 0, 1, 2, or 3 methionines; 0 or 1 histidine; etc.

The complete or nearly complete amino acid sequence of representatives of classes A, B, C, and D can now be compared as illustrated on Fig. 5-8 (522, 352, 144, 524, 359). As expected, there

TABLE 5.2
CLASSIFICATION OF TMV-RELATED STRAINS
BASED ON AMINO ACID COMPOSITION (mole/mole)

Class	*A*	*B*	*C*	*D*	*E*[a]	*F*[a]
Example	*Vulgare*	*Dahlemense*	*U2*	*HR*	*ORSV*	*CV4*
Asp	18	17	22	17	20	17
Thr	16	17	19	13	21	12
Ser	16	16	10	13	12	24
Glu	16	19	16	22	15	10
Pro	8	8	10	8	9	8
Gly	6	6	5	4	7	5
Ala	14	11	17	18	11	20
Val	14	15	12	10	10	14
Met	0	1	2	3	3	0
Ile	9	7	8	8	8	6
Leu	12	13	11	11	14	13
Tyr	4	5	6	7	7	4
Phe	8	8	8	6	7	11
Lys	2	2	1	2	1	3
His	0	0	0	1	0	0
Arg	11	9	8	10	9	10
Cys	1	1	1	1	1	0
Trp	3	3	2	2	3	1
(NH_3)	(19)	(17)	(17+)	(17+)	—	—
Total	158	158	158	156	158	158
Net exchanges from A		8	16	20	17	18
N-terminus		Like A		Like A		
	ac-ser-tyr-		Pro-tyr-			
C-terminus				Like A		
	-pro-ala-thr	-ser	Like A			

[a] Not definitively established until sequenced.

are many more differences than are revealed by the composition analysis, with 27–87 residues being replaced in comparing the amino acid sequences of class A with those of B, C, or D. It should also be noted that in one of these classes (C) the N-terminal acetyl group is lacking, proline being N-terminal (an amino acid that assures enzyme resistance even without acetylation) (359). The proline near the C terminus is present in all four groups. Among other locations or sequences which are common to these four groups are the long segments 87–95 and 113–122, and residues 36–38 (Gln Thr Gln), 27 (cysteine), and 61–63 (Arg Phe Pro). Of the 8 prolines, 5 are immutable and 3 are displaced only in HR (class D). In all respects the latter is farthest removed from the TMV prototype

FIG. 5-8. The amino acid sequence of TMV strains and mutants. V = vulgare = wild type (class A); D (= Dahlemense), U_2 and HR are examples of classes B, C, and D, respectively. Small lettering indicates the location of changes detected after treatment of common TMV (or its RNA) with mutagens, and their frequencies. R (at residues 68 and 139) indicates lysine and tyrosine residues which are chemically reactive in the intact virus (522, 352). Circled residues are those about which there is disagreement between the two laboratories, in four instances regarding the presence of amide groups, and in one instance regarding whether it is threonine or serine (144, 524).

(10) (20) ThrThrMet Ala (30)
V : AcSer-Tyr-Ser-Ile -Thr-Thr-Pro-Ser-Gln-Phe-Val-Phe-Leu-Ser-Ser-Ala-Trp-Ala-Asp-Pro-Ile-Glu-Leu-Ile-Asn-Leu-Cys-Thr-Asn-Ala-Leu-Gly-
Ile (2) His His Leu Met Leu (2) Val Ala Leu(3) Val Val Ser(4) Ile
D : " " " " " Ser " " " " " Phe " " " Val " " " " Ile " " Leu " Val " " Ser-Ser " "
U_2 : Pro " Thr " Asn " " " " " " Tyr " " " Ala-Tyr " " " Val " " Ile " Leu " " Asn-Ala " "
HR : AcSer " Asn " Thr-Asn-Ser-Asn " Tyr- Gln " Phe-Ala-Ala-Val-Trp " Glu " Thr-Pro-Met-Leu " Gln " Val-Ser " " Ser

Lys (40) Gly(2) (50) (60)
V : Asn-Gln-Phe-Gln Thr-Gln-Gln-Ala-Arg-Thr-Val-Val-Gln-Arg-Gln-Phe-Ser-Gln-Val-Trp-Lys-Pro-Ser-Pro-Gln-Val-Thr-Val-Arg -Phe-Pro-Asp
Ser Ala Lys Leu Ala Ile(2) Gly(2) Ser(4)
D : " " " " " " " " " " Thr " " Gln " " Ser-Glu-Val " " " Phe " Gln-Ser " " " " " Gly
U_2 : " " " " " " " " " " Thr " " " " " Ala-Asp-Ala " " " Ser " Val-Met " " " " " Ala
HR : Gln-Ser-Tyr " " " Ala-Gly " Asp " " Arg " " " " Asn-Leu-Leu-Ser-Thr-Ile -Val-Ala- Pro-Asn-Gln " " " Asp

R (70) (80) (90)
V : Ser-Asp-Phe-Lys-Val-Tyr-Arg-Tyr -Asn-Ala-Val-Leu-Asp-Pro-Leu-Val-Thr-Ala-Leu-Leu-Gly-Ala-Phe-Asp-Thr-Arg-Asn-Arg- Ile-Ile-Glu-Val-
Gly(2)Gly Ser (2) Ala(4) Asp
D : Asp-Val-Tyr -Lys " " " " " Ala-Val " " " " Ile " " " " Gly-Thr " " " " " " " " " "
U_2 : Ser-Asp-Phe-Tyr " " " " " Ser-Thr " " " " Ile " " " " Asn-Ser " " " " " " " " Glx "
HR : Thr-Gly-Phe-Arg " " Val-Asn-Ser-Ala- Val- Ile-Lys " " Tyr-Glu " " Met-Lys-Ser " " " " " " " " (Gln)Thr-

(100) (110) (120)
V : Glu-Asn-Gln-Ala-Asn-Pro-Thr-Thr-Ala-Glu-Thr-Leu-Asp-Ala-Thr-Arg-Arg- Val-Asp-Asp-Ala-Thr-Val-Ala-Ile-Arg-Ser-Ala-Ile-Asn-Asn-Leu-
Gly (2) Arg Met(3) Cys Gly Val Ser
D : " " Gln--Gln-Ser "
U_2 : Asx-Asx-Glx-Ala-Asx " " " " (Asx,Thr,Glx,Glx, Pro, Val,Ile) " " " " " " " " " " Ala-Ser " " " "
HR : Glu-Glu- GlnSer-Arg " Ser-Ala-Ser-Gln-Val -Ala(Asn)Ala -Thr-Gln " " " " " " " " " " Ser-Gln " Gln-Leu "

(130) R (140) (150)
V : Ile-Val-Glu-Leu-Ile-Arg-Gly-Thr-Gly-Ser-Tyr-Asn-Arg-Ser-Ser-Phe-Glu-Ser-Ser-Ser-Gly-Leu-Val-Trp-Thr-Ser-Gly-Pro-Ala-Thr-
Thr Val Gly(2) Ile Phe(3) Cys Lys Phe Ser Ile Leu(3)
D : Val-Asn " " Val " " " " Leu Tyr " Gln-Asn-Thr " " Ser- Met " " " " " " " Ala " " Ser
U_2 : Ala-Asn " " Val " " " " Met-Phe " Gln- Ala-Gly " " Thr-Ala " " " " " " " -Thr " " Thr
HR : Leu " " "Ser-Asn-His-Gly " Tyr-Met(")-Arg- " (Glu) " " — Ala-Ile — " Pro " (") " Ala " " "

and is the only TMV strain to show a different number of amino acids, 156 rather than 158. As judged from sequence similarities, the deletion of two amino acids must have occurred near the C terminus (146–149), a part of the molecule near the virus surface and therefore more mutable. The two long immutable regions are boxed in on Fig. 5-7. We shall make further references to the coat protein differences of the TMV strains in later chapters.

2. *Other Plant Virus Proteins*

Concerning the proteins of other plant viruses, the terminal sequences and extensive though incomplete internal sequence data are available for the TYMV protein (168). Terminal amino acids or sequences are also known for many other plant viruses, as reviewed elsewhere (122). Almost all have nondetectable N termini, which usually are presumed to be acetylated. Most of the subunit molecular weights range up to 20,000, and when they seem to be higher the possibility of aggregates or dimers has not always been completely excluded (308).

One remarkable instance of an incomplete virus has attracted considerable attention and interest in recent years, and it is discussed in greater detail later. It is the satellite of tobacco necrosis virus (STNV), a small isometric particle which can replicate only on inoculation together with the larger and immunologically unrelated tobacco necrosis virus (TNV) (see Section VIII,F). The protein of the satellite (STNV) has been reported to have a minimal molecular weight, on the basis of its cysteine content, of 13,500. The amounts of free N-terminal and C-terminal alanine found suggest a molecular weight of 39,000, but recent data favor a molecular weight of 27,000 (351, 367).

3. *Animal Virus Proteins*

Most animal viruses occur at very low concentrations in the infected cells (less than 1%), and their scarcity has greatly impeded their chemical study. Several of the animal picorna viruses [poliomyelitis, mouse encephalitis (ME), encephalomyocarditis (EMC)] show considerable similarities, including the fact that they contain two or three major and one or two minor protein components in their capsid (a total of four, as revealed by gel electrophoresis) which also differ in their amino acid composition (282, 284). In the case of the ME virus, it was clearly established that these components were not produced by the reducing agent from a single capsid protein consisting of several disulfide-linked peptide chains (373, 374). The molecular weight of the protein components of the polio virus are 35,000, 28,000, 24,000, and 5000, and that of the major components of the ME virus near 26,000.

The foot and mouth disease virus seems to contain a single protein the molecular weight of which, as derived from carboxypeptidase action, would be 66,000, although a lower value is regarded as more probable (20).

Considerable work has been done on the multiple protein components of the most interesting of the myxoviruses, the influenza virus and its strains. It appears that the nucleocapsid protein of this virus (38% of the total protein) has a molecular weight of about 40,000 (261). Quite similar peptide patterns are obtained on tryptic digestion of this protein as derived from different strains. In contrast, the hemagglutinin (37%) derived from its envelope varies greatly from strain to strain. It seems to be composed of peptide chains with N-terminal aspartic acid, averaging 60,000 in molecular weight on that basis. Least is known about the chemical nature of the viral neuraminidase. Its sedimentation rate is much higher than that of bacteria or chick embryo cells (9 S vs 5.5 and 3 S) (501, 86). The location of these proteins in the virus particle and their biological roles are discussed later (see Chapter VII).

Little is known about the proteins of the paramyxoviruses except that they too contain neuraminidase and hemagglutinin. The reoviruses seem to consist of three major and four minor proteins, but their nucleocapsids are composed of only two proteins (275). In contrast, the Sindbis virus, an encephalovirus, appears to consist of only one lipid-rich envelope protein and one capsid protein (441, 550). The RNA tumor viruses which have been purified [Rous sarcoma virus, RSV, and the related viruses, RSV (RAV), RSV(O), RSV(SR-RSV), and avian myeloblastosis] show the same two major and one minor proteins, as judged by gel electrophoresis under two different conditions of solvent and pH (Fig. 5-9) (89). The major components could be identified with the internal group-specific antigens which are common to those viruses and apparently represent the nucleocapsid proteins, since the group-specific antibody

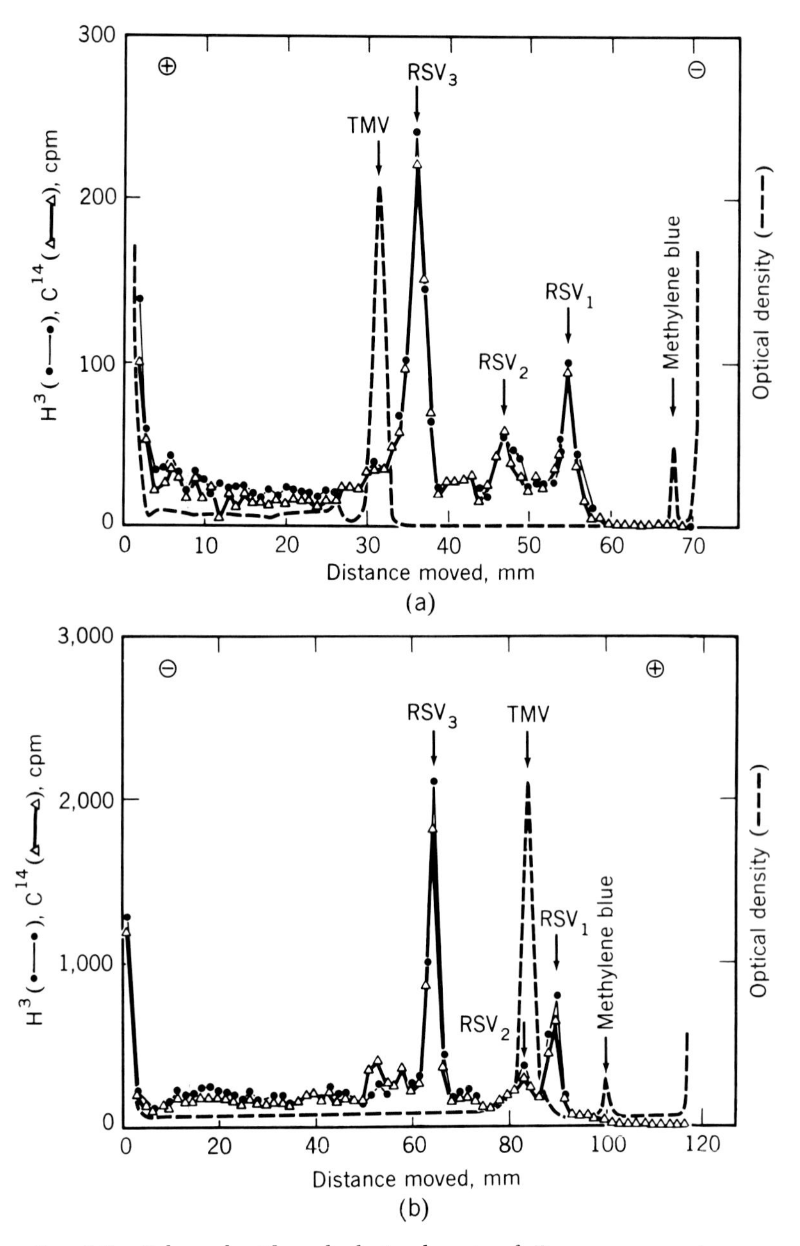

FIG. 5-9. Polyacrylamide gel electrophoresis of Rous sarcoma virus proteins. (*a*) Migration in 8 M urea (pH 3.8); (*b*) in 0.1% SDS (pH 7.2). TMV protein is added as a UV absorbing marker in both. The Rous proteins are detected by their radioactivity (89).

does not neutralize or inactivate the intact viruses (see Chapter XII).*

Of the proteins of the animal DNA viruses, those of the adenovirus group have been studied most intensely. Of the ten protein components of this virus of 177×10^6 daltons, the three main components of the outer shells, the hexon, penton, and penton fiber (see Section VIII,C), have been separated (476), and the hexon and fiber of types 2 and 5 have been analyzed for their composition (336). The weight of the hexon and penton were estimated as 210,000 daltons, but more recent data favor about 360,000 (285, 286). The fiber, including the knob at the end (see Section VIII,C), seems to be a protein molecule of 70,000 molecular weight. Besides these structural proteins of the outer shell, there also occur several internal capsid proteins amounting to 20% and resembling histones and protamines in their high content of basic amino acids (377). Like histones, they have alanine as the predominant N-terminal amino acid (263).

The Shope papilloma virus which may contain one predominant protein component has been analyzed for end groups. No N-terminal residue was detected, but a C-terminal sequence-(Tyr, Leu)-Ala-Thr corresponding to a molecular weight of 160,000 was reported (240).

Quite intense studies are now concerned with the polyomavirus and other small DNA viruses of similar properties. The polyoma protein appears to have a molecular weight of about 44,000 of typical composition (113, 464). Related to polyoma is the SV40 virus. This simian virus contains two main protein components of different charge (A and B, amounting to 91%) which form the virus shell, and a basic third component which forms a complex with the DNA (12, 13, 253). The amino acid compositions and molecular weights (all about 16,700) were recently reported (388).

4. *Bacteriophage Proteins*

The proteins of the bacterial viruses have been studied in more detail than those of the animal viruses. The only virus proteins,

* A very similar protein pattern was obtained also with the Friend mouse leukemia virus (539).

besides those of TMV and its strains, which have been completely sequenced are the coat proteins of several small bacteriophages. Different members of a group of RNA phages show more or less close relationships to one another, resembling in their range of serological differences those among the various TMV strains (397). Two representative types containing 129 amino acids have been sequenced (500, 521, 270). A considerable number of phages (f2, MS2, M12, R17) closely resemble one another, since they differ only by one or two amino acids substitutions. In contrast, fr is a more distant relative, with about 20 replacements. Nevertheless, fr and the f2 group have the same number of amino acids, and long segments of the two peptide chains are identical (residues 7–16, 28–53, 70–85), as are the locations of all prolines as well as of most of the larger amino acid residues (Trp, Tyr, Phe, Leu, Met, Arg, Lys). Both types lack histidine, and both have two cysteine residues in the same locations Fig. 5-10).

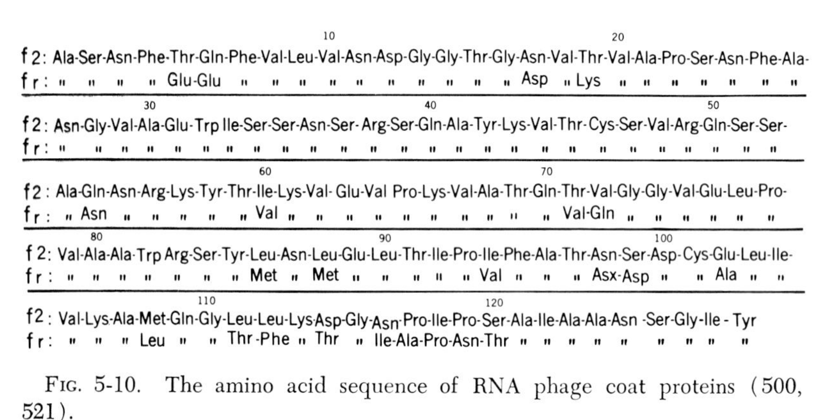

FIG. 5-10. The amino acid sequence of RNA phage coat proteins (500, 521).

In contrast to the TMV strains, the N-termini of which are acetylated, or, in one instance, proline, the N-terminus of MS2 and related phages is nonacylated alanine. However, this chain end is hidden, presumably through the folding of the peptide chain, and it reacts with typical amino group reagents only very incompletely, even in 8 *M* urea, and not quantitatively in 6 *M* guanidine hydrochloride. The C-terminus of the peptide chain of the phages is susceptible to limited digestion by carboxypeptidase, but in the intact phage, unlike TMV, the phage protein is enzyme-resistant (270).

In addition to the coat protein, the RNA phages of the f2 group contain one molecule of another protein, the maturation factor or A protein, of molecular weight 38,000, and containing five histidines (437, 438). This amino acid is absent in the coat protein.

One other virus protein which has been subjected to amino acid sequence analysis is the coat protein of fd, a small rod-shaped virus containing DNA, the extraordinary properties of which are discussed later (45). The fd coat protein is by far the smallest structural protein known, consisting of only 49 amino acids, and lacking arginine, histidine, and cyst(e)ine. Its sequence is:

```
Ala-Glu-Gly-Asp-Asp-Pro-Ala-Lys-Ala-Ala-
                5                 10
Phe-Asp-Ser-Leu-Glu-Ala-Ser-Ala-Thr-Glu-
                15                20
Tyr-Ile-Gly-Tyr-(Ala, Trp, Gly, Val, Val, Val,
                (25)                  (30)
Val, Met, Ile)-Ala-Thr-Ile-Gly-Ile-Lys-Leu-
                35                40
Phe-Lys-Lys-Phe-Thr-Ser-Lys-Ala-Ser
                45            49
```

Mutants differing in composition have been isolated electrophoretically. Similarities in the C-terminal segment of this protein with that of the TMV family, all rod-shaped viruses, have been pointed out by Braunitzer. Instead of the lysine as the third residue of fd, there is Pro-Gly or Pro-Ala in TMV and strains.

Evidence has recently been obtained by controlled alkali degradation that in f1, a related phage (534), there exists a second capsid protein containing arginine and histidine, which plays a role in the attachment of the tip of the phage rods to the pili, possibly an equivalent of the A-protein of the RNA phages (361).

The other type of small DNA bacteriophages, exemplified by ϕX174, contains three major and one minor component, all differing in composition and electrophoretic mobility in 9 M urea (pH 2.3) (425). The two major components have one N-terminal Ser-Asp- and Met-Ser- sequence, respectively, per 25,000 molecular weight, both in ϕX174 and in S13 (425, 341). Several of these proteins have been tentatively allocated to various cistrons and to particular structural features of those viruses (e.g., the spikes which are es-

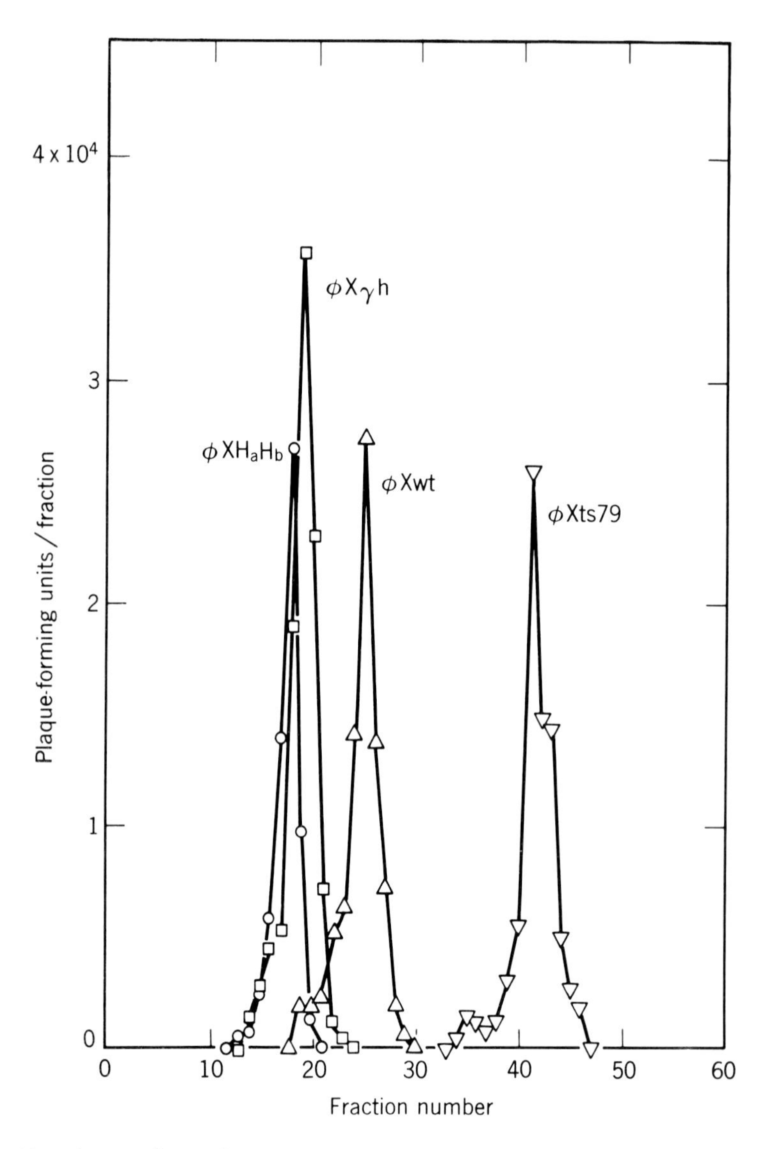

FIG. 5-11. Electrophoretic separation of four strains of φX174. Electrophoresis on polyacrylamide gel (2.6%) in pH 9.5 buffer (425).

sential for adsorption of the virus by the host cell) (see Fig. 5-11).

Concerning the proteins of the medium-sized phage, λ, two components of different composition and molecular weights 53,000 and 110,000 have been identified (93).

The complexity of structure of the large T-even bacteriophages makes it a foregone conclusion that several, if not many, structural proteins exist. The main morphological component, the head, accounts for over 85% of the protein and is now believed to be composed of subunits of molecular weight 42,000 with N-terminal alanine and C-terminal glycine (74, 103, 258).

The other structural components of the T-even phages, all parts of the tail, have no detectable N-terminal residues (46). The contractile sheath contains 144 electronmicroscopically visible subunits and 144 histidine residues. Two C-terminal amino acids were detected, by both hydrazinolysis and carboxypeptidase (glycine and serine, 3:1), and they amounted to about 280 per sheath (258). Thus the presence of at least two proteins is indicated. Distinct differences in amino acid composition were noted for whole phage, contractile sheath, and tail tubes.

The complete primary structure of the lysozyme of T2 and T4 has recently been reported. These two proteins contain 164 residues and differ by only three amino acid replacements ($Asn_{40} \rightarrow Ser$, $Ala_{41} \rightarrow Val$, $Thr_{151} \rightarrow Ala$, in T4 vs T2). Surprisingly, there is no resemblance between these phage lysozymes and egg white lysozyme, as illustrated by the presence of two cysteines and three tryptophans in the phage enzymes, as compared to four cysteines and six tryptophans in the vertebrate enzyme. Yet, functionally they at least resemble one another (216a, 473a).

CHAPTER VI

THE NUCLEIC ACIDS OF VIRUSES

A. Chemical Structure

1. *Components*

Nucleic acids are polymers composed of nucleotides, one molecule of water being lost in the formation of each internucleotide phosphoester bond. The nucleotides consist of three types of components: a heterocyclic "base," a sugar, and phosphoric acid. Typical RNA contains the two purines, adenine and guanine, and the two pyrimidines, cytosine and uracil; in DNA the uracil is replaced by thymine. The sugar is *r*ibose in *R*NA (*r*ibonucleic acid) and *d*eoxyribose in *D*NA (*d*eoxyribonucleic acid). A glycosidic bond links the —NH group in the 9 position of the purines and that in the 1 position of the pyrimidines to the aldehyde (1′) group of the sugar. These sugar-base compounds are called adenosine, guanosine, cytidine, uridine, and thymidine, the prefix "deoxy" being used for the deoxyribose derivatives. The phosphoric acid is esterified with the 3′- and 5′—OH groups of the sugar of neighboring nucleotides in nucleic acid. Different enzymatic methods of degradation can leave the phosphate groups attached to either the 3′ or the 5′ position of the sugar, yielding the 3′- and 5′-nucleotides or -deoxynucleotides (generally symbolized by Xp and pX) respectively (see Fig. 6-1).

The 2′—OH group of the ribonucleotides greatly increases the alkali sensitivity of the 5′-ester linkage of the phosphate in the 3′ position, since cyclic 2′-, 3′-phosphates tend to form at the expense of the 5′-linkage. Nucleic acids are frequently symbolized as shown in Fig. 6-2, which illustrates the effect of alkali and of snake venom diesterase on a polyribonucleotide. When RNA is degraded by alkali, the final reaction product is a mixture of 2′- and 3′-nucleotides. Ribonucleases, like alkali, give intermediate cyclic phosphates (see Fig. 6-1), but the final product consists of 3′-nucleotides only.

In contrast to the alkali lability of RNA, DNA, lacking the 2′-OH group, is quite alkali-resistant. However, DNA is particularly

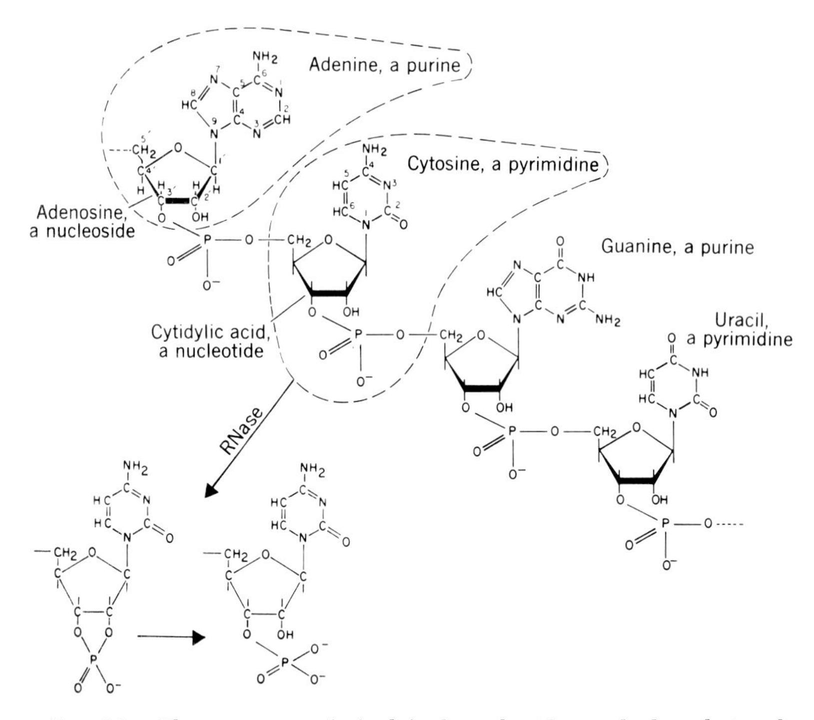

Fig. 6-1. The structure of (poly)-ribonucleotides and degradation by pancreatic ribonuclease. Note the numbering system detailed for adenosine and cytosine.

susceptible to acid because the glycosidic bond (between the sugar and the ring compound) is quite unstable in deoxynucleotides. The purine-deoxyribose bond begins to break at ambient temperatures below pH 4, whereas rupture of the pyrimidine deoxyriboside and

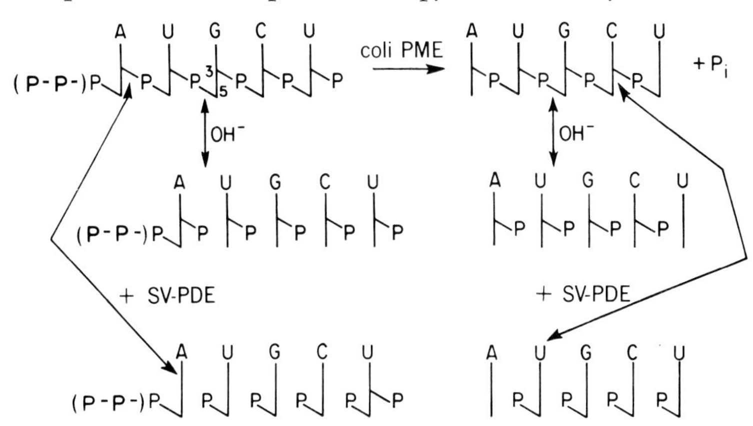

FIG. 6-2. Terminal dephosphorylation and degradation of oligo- (or poly)-nucleotides by alkali and snake venom diesterase. Note that the methods, singly or in combination, make identification of terminal residues possible, since these can be obtained with one or several ionized phosphate groups more or less than the bulk of the internal nucleotides. (Coli PME = phosphatase; SV-PDE = snake venom phosphodiesterase; OH— = M KOH; all used at 25°.)

the purine riboside bonds require heating in acid. The pyrimidine ribosides (or ribotides) are not appreciably degraded at 100° in 1 *N* HCl, but all bases can be released in much stronger acid and /or at much higher temperatures. The ribose- and deoxyribose-phosphate bonds represent the chemically most stable part of the nucleotides. Thus the nucleosides and deoxynucleosides are usually not prepared by chemical means but by enzymatic dephosphorylation of nucleotides.

The so-called bases are not ionized under physiological conditions. The pK_a values for the amino groups on guanosine, adenosine, and cytidine (the negative logarithm of their acidity constant) are about 2.7, 3.5, and 4.2; this means that at these pH's the amino groups are approximately 50% protonated ($—NH_2 \rightarrow NH_3^+$). The oxygen group of guanosine and uridine undergoes a tautomeric shift and dissociation near pH 9.5.

$$-NH-C\overset{O}{\diagup\!\!\diagup}\diagdown \longrightarrow \left(-N=C\overset{OH}{\diagup}\diagdown \longrightarrow\right) -N=C\overset{O^-}{\diagup}\diagdown \quad H^+$$

The general nature of the bases, and particularly of the purines, is hydrophobic, and only the ribose moiety and the ionized phosphates render nucleic acids hydrophilic and water-soluble. The pK_a of the doubly esterified phosphate is about 1; that is, all phosphates in RNA are fully ionized, except for the secondary dissociation at terminal phosphates which has a pK_a of about 6. The fact that terminally phosphorylated oligonucleotides lose one charge between pH 7 and 5 is utilized in procedures for their separation.

2. *End Groups*

As usual, when the chemical nature of chain polymers is under study, end group analyses represent the opening wedge in their elucidation. The nucleic acids can carry either one or several phosphate groups on one or the other terminal —OH group (5′ or 3′), or these groups can be unphosphorylated. If terminal phosphate groups are present, they can be removed by the purified *E. coli* alkaline phosphatase, and determined as inorganic phosphate. A method of degradation can then be selected that yields one or the other terminal residue in a state different from all internal nucleotides. Thus the end group can be isolated as the nucleoside or as the nucleoside di-, tri-, or tetraphosphate, while the rest of the molecule has been reduced to 5′- or 3′-nucleotides (see Fig. 6-2). These methods were first applied to TMV RNA and have since been used with various other viral nucleic acids. Unphosphorylated adenosine was isolated from both ends of TMV RNA (449, 450), and the absence of terminal residues in ϕX174 DNA presented the first evidence that this molecule was cyclic (110).

3. *Terminal Sequences*

Two paths are available for further structural elucidation of viral nucleic acid, namely, stepwise degradation and complete degradation by specific nucleases to oligonucleotides.

A method for stepwise degradation of viral RNA has been developed and successfully applied to identify the last five nucleotides on the right (5′- or 3′-linked) end of the TMV RNA molecule (Fig. 6-3) (435, 436). There is no reason why this method should not be refined to a point where 10 or 20 nucleotides could be released in stepwise fashion and identified. The method consists in oxidizing the terminal glycol group (two neighboring —OH groups in the *cis* position) to two aldehyde groups. The aldehyde group in the 3′ position

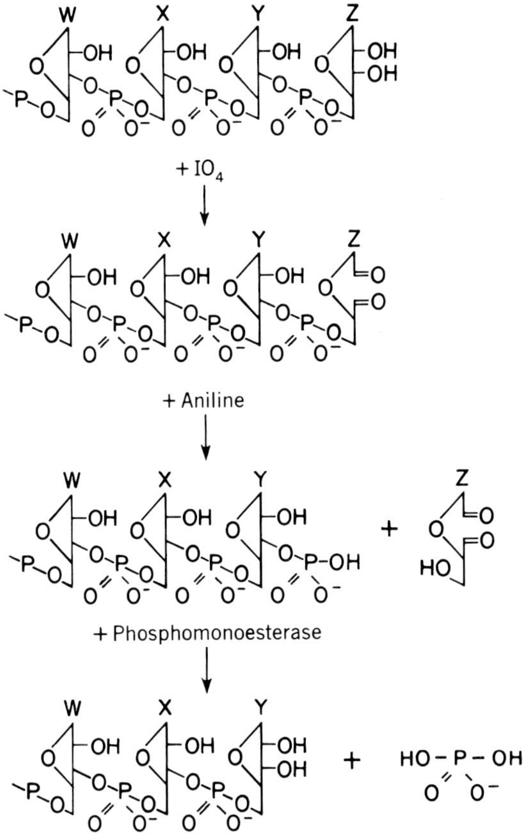

FIG. 6-3. Stepwise degradation of RNA. All steps performed in the range of pH 5–7 at 0–25° (see text) (436).

weakens the ester bond on the β-carbon (5′) so that the oxidized terminal nucleoside is released at ambient temperature under the catalytic influence of aniline at pH 5. The residual polynucleotide chain, now terminally phosphorylated, is treated with phosphatase, thus releasing the phosphate since this enzyme is specific for mono-esterified phosphates. A new glycol group is generated, and these cycles of oxidation, β-elimination, and dephosphorylation can be repeated until the cumulative effect of the many steps leads to some chain breakage and thus to additional spurious end groups and unclear results. The released oxidized nucleoside (only a single one at each step under ideal conditions), readily yields the free base on acid treatment. The base can then be identified chromatographically.*

This and all other methods of identifying one or a few of the 3000 to 200,000 nucleotides making up a viral nucleic acid molecule require enormous amounts of the precious materials, unless use is made of radioactive labels in the nucleic acid or in the reagents employed to modify the nucleic acid. For example, by holding infected plants in a $^{14}CO_2$-containing atmosphere, plant viruses and their RNA giving 1000–5000 cpm/μg can be obtained and, using such material, the terminal base can be detected and identified in about 1 mg of viral RNA.

Another approach to structural analysis of viral RNA makes use of either of the two nucleases now known to have definite specificities. Pancreatic ribonuclease (RNase) splits only on the right of the pyrimidine phosphate (see Fig. 6–1), and the T1 nuclease from takadiastase only on the right of guanine residues. All fragments resulting from pancreatic RNase contain only one pyrimidine which is terminal (Cp; Up; XpCp; XpUp; XpXpCp; XpXpUp; etc.), and all T1 fragments have only one terminal guanine (Gp; XpGp; XpXpGp; XpXpXpGp, etc.). The fragment derived from the right end of the RNA chain, if terminating in unphosphorylated A as it does in TMV RNA and the f2 class of the RNA phages, is the only fragment obviously lacking a terminal pyrimidine or guanine if it is isolated after pancreatic or T1 RNase digestion, respectively. More

* TMV RNA loses 85% of its infectivity upon oxidation and over 95% upon removal of the terminal A (436). In contrast, R17 phage RNA is inactivated only upon removal of the second nucleotide (C) (541). The terminal A is restored in the progeny of the terminally de-adenylated RNA.

significant from the point of view of isolating the terminal fragments from the grand mixture is that they, like the terminal residues obtained by complete degradation, often differ in phosphate charges from all others. Thus, on degradation by T1 ribonuclease, the right end of TMV RNA yields only one oligonucleotide fragment lacking a 3′-phosphomonoester group. Its charge is therefore the same at PH5 and 7.5, and it is unaffected by phosphatase, which in all other fragments causes the loss of 2 negative charges. Similarly, on pancreatic or T1 ribonuclease treatment, MS2 RNA, which starts at the left with a triphosphate group, yields only one fragment which loses 6 negative charges on phosphatase treatment, all others losing only 2 charges. (The *E. coli* phosphatase slowly splits the interphosphate bonds, pppX → ppX → pX → X, only the last step being fast.)

A very useful research tool to exploit these facts is the separation of oligonucleotides on DEAE cellulose or DEAE Sephadex columns, since these polycationic resins retard anionic materials in proportion to their negative charges. That nucleotides tend to interact in solution and to form larger complexes frequently obscures this law. However, the discovery that the presence of 7 *M* urea in the eluting buffers interferes with nucleotide complexing has made the separation of nucleotide fragments according to number of charges a very valuable research tool (467).

The isolation of terminal fragments has become possible on the basis of the principles discussed above. These procedures are responsible for the identification of the right terminal undecanucleotide (11 residues) (150, 494) and of the left terminal trinucleotide in the MS2 class of the RNA phages (149), as well as confirmation of the right terminal pentanucleotide (5 residues) sequence in TMV RNA (150, 289) (Table 6.1).

This type of study is further facilitated by attaching a radioactive label to the end of the chain. One procedure now extensively used is to subject the RNA to the same type of terminal oxidation by periodate as used in stepwise degradation (see Fig. 6-3), and then to reduce the dialdehyde with ^{3}H-containing sodium borohydride ($NaBH_4$) (346). Alternatively, the dialdehyde can be condensed with typical aldehyde reagents, such as ^{14}C-labeled semicarbazide or nicotinic acid hydrazide (435, 212). These reactions are more stoichiometric and selective in regard to end-group

TABLE 6.1

STRUCTURAL FEATURES OF VIRAL NUCLEIC ACIDS[a]

Virus	*3′-Linked end*[a]		*Internal nucleotides*			*5′-Linked end*[a]	*References*
TMV[b]	A(*U–G*)~1500	LLS[b]	~1800	ψ2	~2500	ψ1, ~120, ω, ~120-*G-C-C-C-A*	449, 450, 435, 436, 150, 289, 290
R17, MS2, f2	*pppG-G-U-*[d]		~3000			-*G-U-U-A-C-C-A-C-C-C-A*	149, 150, 358a
Qβ[c]	(a) pppG-G-G-G-A-A-C-[d]		~2800			-G-C-C-C-U-C-U-C-U-C-U-C-C-C-C-A	35a, 74a, 509, 552 536a
STN	ppA-G-U		~1200			G(U, A, A, C)C[d]	517, 553, 554
TYMV	A-Py-						456

[a] Italicized sequences have been determined by different methods and/or investigators. Minority opinions are not included. The numbers represent approximate numbers of nucleotides.

[b] LLS stands for the local lesion gene on *N. sylvestris* and ψ1, ψ2, and ω for G lacking nucleotide sequences of about 26 and 70 nucleotides.

[c] A shorter variant of *in vivo* synthesized Qβ RNA has also given the 5′ terminal octanucleotide sequence indicated. *In vivo* synthesized Qβ terminates in -C, not in the sequence given above. (35b)

[d] Other sequences reported to occur at the 3′-linked end f2 and Qβ RNA (494, 552) and the 5′-linked end of STN RNA (553), contain one more G and C, respectively, then listed above.

labeling than the tritium reduction, but the aldehyde-amine compounds are not sufficiently stable for many contemplated uses of such labeled RNA.

Another useful label is obtained by introduction of a phosphate group on the unphosphorylated (or enzymatically dephosphorylated) left terminal 5′—OH group by the enzyme polynucleotide phosphokinase (from phage-infected *E. coli*) which with great specificity transfers phosphate from ATP to such —OH groups (353). Thus, with γ ^{32}P-labeled ATP (p*-p-p-A) as donor, that label and two additional negative charges can be attached to the left end of viral RNA prior to complete or selective degradation and separation of fragments.

Several enzymes are known to degrade nucleic acids sequentially, starting at one end or the other. They have proved very useful in establishing the sequence of oligonucleotides, and they can also be used to advantage if an RNA is to be degraded completely into 3′-nucleotides (spleen exonuclease, attacking nucleic acids from the 5′-end) or into 5′-phosphorylated nucleosides (snake venom phosphodiesterase, giving pX, or polynucleotide phosphorylase, ppX, both enzymes starting from the 3′-end). However, attempts to use this type of enzyme for sequence analysis of nucleic acids as large as viral RNA's in a manner analogous to the use of aminopeptidases and carboxypeptidases for protein sequence analysis have yielded misleading results. It should also be noted that unmodified unsubstituted end groups are greatly favored by these exonucleases. Thus spleen exonuclease does not attack 5′-phosphorylated ends, and the snake venom enzyme is greatly inhibited if the terminal glycol group has been oxidized and reduced (149). An exonuclease that has been successfully used with viral nucleic acid is the phosphodiesterase from *E. coli* (see Section VI,D) (110).

4. *Internal Nucleotide Sequences*

The methods of selective endonuclease degradation and oligonucleotide fractionation (see Section VI,A,3) performed for the purpose of obtaining the terminal fragments also yield material and data pertaining to the ultimate complete structural elucidation of viral nucleic acids. Thus, on ribonuclease degradation and separation according to chain length, each nucleic acid gives a characteristic pattern of different amounts of mono-, di-, tri-, tetra-nucleotides, etc., of all possible combinations. These size groups (isopliths) can be further resolved by other chromatographic and by electrophoretic methods (375, 462, 463). However, longer fragments of 8–12 purines terminated by a pyrimidine are sufficiently rare for there to exist only a limited and characteristic number (0–4) of each such purine run in the RNA of a given virus (423, 545a). Particularly from T1 digests unique fragments of considerable magnitude (containing only one terminal G) have been isolated and are being subjected to sequence analysis. It now appears that in TMV RNA there are two runs ("uniquemers") of 26 nucleotides, and one of 70, lacking G (Table 6.1; see also Section VIII, J) (290). The recognition of such unique fragments may well be of considerable usefulness in advancing understanding of the structure and function of viral nucleic acids.

B. Conformation

The planar and hydrophobic nature of the bases gives them a tendency to interact in polynucleotides. This so-called *stacking of bases* plays a distinct role in the chemical and physical properties of single stranded nucleic acids and it also favors, and contributes to, the interaction of complementary bases which is the basis of double strandedness.

The concept of *complementary bases*, the principal biological and chemical fact concerning nucleic acids, states that each of the bases as present in a polynucleotide has a high binding affinity for another (498). The two resulting pairs are the 6-aminopurine, adenine, and the 4-oxopyrimidine, uracil (or thymine) on the one hand; and the 6-oxopurine, guanine and the 4-aminopyrimidine, cytosine, on the other. As illustrated, the A-T and A-U pairs are held together by 2 hydrogen bonds, and the G-C pair by 3. (Fig. 6-4). The strainless fitting of two polynucleotide strands requires a helical antiparallel conformation.

The simplest illustration of double strandedness is the effect of mixing salt containing solutions of polymerized adenylic and uridylic

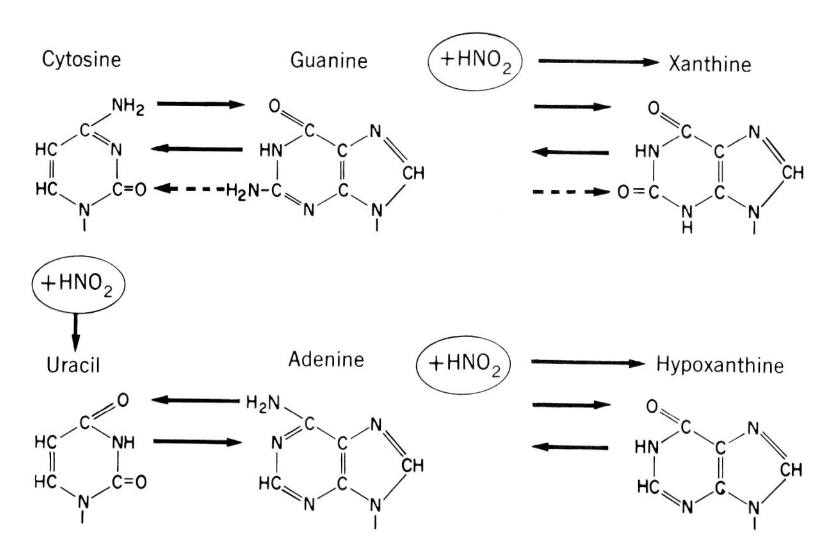

FIG. 6-4. Complementarity of bases, A-U (or T) and G-C. The hydrogen-bonded interaction of complementary base pairs is illustrated by arrows (from the H-bond donor to the acceptor). The action of nitrous acid in changing the direction of arrows in C and A is discussed in Chapter IX.

acids, respectively (poly A and poly U). If the solutions are used in equimolar proportion, a double stranded polymer is formed spontaneously which shows the following distinctive properties, resulting from the pairing of each A residue with a U residue: (a) the complex has only about 60% of the UV absorbance calculated for the mixture; (b) the complex shows much higher optical rotation (—300° as compared to about —30° for the single stranded polynucleotides); (c) the complex lacks the chemical reactivity at the -N-1 and at the amino group of A, (d) the complex is resistant to nucleases. Although a hydrogen bond has only about 4 k cal, the cumulative effect of hundreds of residues interacting by as many bonds renders poly A-poly U a very stable double stranded compound.

Because natural nucleic acids are rarely as simple in composition as poly A, it would be surmised *a priori* that for a natural nucleic acid to encounter a nucleic acid with complementary antiparallel nucleotide sequence would be a highly improbable event. This is not so, because the mechanism of biosynthesis of nucleic acids is actually one of complementary replication. Any nucleic acid can serve as template for an appropriate polymerase (or replicase). The four 5′-triphosphates are the required building blocks, and they become interconnected by 3′–5′ monophosphate bridges, pyrophosphate being released. The growth of the complementary chain proceeds from the 3′- to the 5′-end of the template, thus producing the antiparallel chain starting with pppX- (Fig. 6-5). The nucleo-

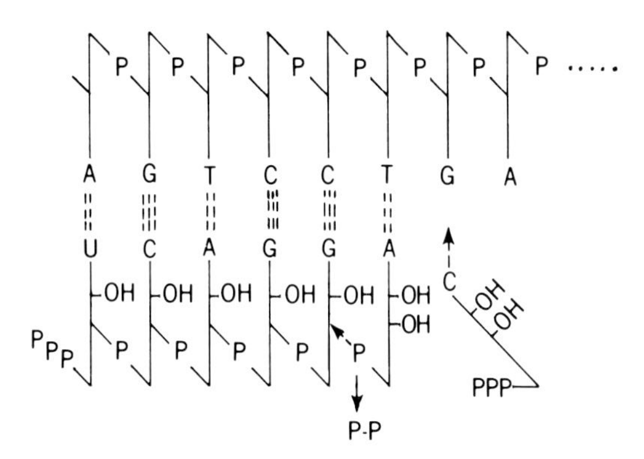

FIG. 6-5. Transcription. The template action of a polynucleotide in the alignment of triphosphates for polymerization is illustrated with a DNA chain as template and RNA as product. The same base-pairing principle operates for replication of DNA or RNA, as well as for translation (see Fig. 6-8).

tide sequence of the newly synthesized chain is not determined by the enzyme but by the template, and the product of this reaction is double stranded nucleic acid, RNA or DNA, as the case may be (see Chapter IX, B). In view of this biosynthetic mechanism, it is not the existence of complementary double stranded nucleic acid that appears surprising, but rather the occurrence of nucleic acids in nature also in single stranded form. Special biological mechanisms must be invoked to account for their existence.

In the laboratory double stranded DNA in salt-containing solution (e.g., 0.15 *M* NaCl, 0.015 *M* sodium citrate) must be heated to 90–100°, depending on its G + C content, and quickly cooled to obtain largely single stranded DNA, as indicated by changes in the various criteria listed above; this process is termed melting (Fig. 6-6). Double stranded RNA is even more stable. However, at

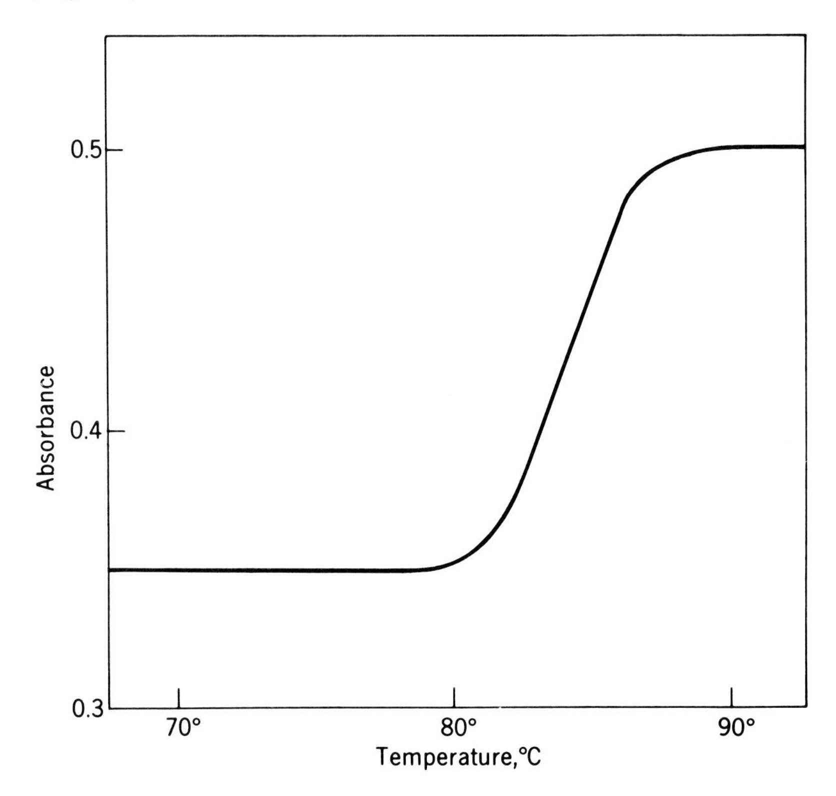

FIG. 6-6. Melting of DNA. The typical effect of heating double stranded nucleic acids in the presence of salt (0.15 *M*) on their UV absorbance.

low ionic strength the mutual repulsion of the many negative phosphate charges along the chain becomes a major factor in weakening the helix structure, and the conformation tends to disrupt at lower temperatures. It seems paradoxical, but it is true that double stranded DNA may be heated to near boiling in salt but is decomposed at room temperature when in distilled water. Other agents which lower the melting temperature of double stranded nucleic acids are 8 *M* urea, 90% formamide or dimethylformamide, and, particularly, dimethylsulfoxide (>80%), the latter causing complete dissociation in the cold even in the presence of salts.

The words "quickly cooled" in the preceding description of the method for obtaining single stranded or "denatured" DNA are significant. When complementary polynucleotide strands are held for some time at elevated temperatures below their characteristic melting temperature, they gradually return to double strandedness (the state utilizing their bonding capabilities most fully) and thus to their most stable configuration. This "*renaturation*" (291) has proved to be a powerful tool in detecting identical (or very similar) nucleotide sequences in various nucleic acids by the so-called hybridization techniques. Also, that certain double stranded nucleic acids renature even on quick cooling and in dilute solution has been an important indication that interstrand crosslinking can prevent the denatured strands from separating. Even more significant is the same behavior in the case of cyclic nucleic acids, because it shows that the rings are intertwined (catenated) and cannot be separated, *in vitro* or *in vivo,* without opening one of the rings. See Section VI, D.

C. General Features of Viral Nuceic Acids

Most viruses contain one molecule of nucleic acid per particle. As is shown later, this does not preclude that these molecules differ from one particle to another, as exemplified by the T-even phages, but generally the viral genome exists as one chemical entity, and occasionally as several definite chemical entities.

In almost all plant viruses the nucleic acid is RNA,*. In animal and bacterial viruses it can be either RNA or DNA. RNA usually occurs as a single stranded chain, but at least two plant viruses

* Cauliflower mosaic virus was recently shown to contain infective DNA (407). The phagelike DNA viruses of blue-green algae (e.g., LPP-1) (389) represent a special taxonomic problem, since these algae resemble bacteria in other respects also.

(wound tumor and rice dwarf) and one group of animal viruses (reovirus) contain double stranded RNA.

Several small viruses of plants, animals, and bacteria contain single stranded RNA of about 10^6 molecular weight, and another group RNA of about 2×10^6 (the picorna viruses). Yet in the first group the sedimentation rate ranges from 17 to 30 S, and in the second group from 30 to 36 S. The higher values are attributed to a more compact conformation, as is the comparatively small radius of gyration (about 20 mμ) in 0.1 *M* salt. Differences in sensitivity to nucleases may also be a consequence of such differences in conformation. The more compact conformation of the RNA of the Qβ and f2 phage groups is probably retained from its intraviral state, since treatment with EDTA seems to abolish it.

The very small DNA phages and viruses, fd, φX174, and MMV, contain single stranded DNA, but in all medium-sized and large DNA viruses the DNA is double stranded. The DNA of the very small phages is not only single stranded, but also cyclic, and it now appears probable that these molecules are truly circular, with no unusual linkers, and no bonds other than 3′–5′ phosphodiester bonds. This is particularly unexpected for the fd type phages, since they are long, fibrous particles (see Table 3.1 and Fig. 11-1).

Circularity has also been found to be characteristic of several other classes of double stranded DNA, both viral and cellular (procaryotic and eucaryotic). Not always, as shown later, are the double stranded circles closed by primary linkages.

In view of the wide range of structural possibilities, the characterization of a viral nucleic acid must begin by establishing whether it is (a) RNA or DNA, (b) single or double stranded, (c) linear or circular, and (d) homogeneous or heterogeneous under conditions that favor molecular disaggregation.

(a) If the nucleic acid is degraded by 0.1 *N* alkali at 25–37° and if it contains uracil rather than thymine (the two easiest tests), it is RNA. If purines are largely liberated at pH 2-3, it is DNA. Color tests are also available to differentiate ribose from deoxyribose (the orcinol test, and the diphenylamine reaction), but these are not fashionable methods.

(b) If the base analysis [nucleotide analysis after alkali degradation, if RNA; base analysis after performic acid degradation (12

N, 1 hr, 175°) for RNA or DNA] indicates approximately equal values for A and T or U on the one hand, and for C and G on the other, then double strandedness is suggested. The UV absorbancy of single stranded nucleic acid is markedly affected by the presence of salts. The RNA is hypochromic in 0.1 *M* NaCl or 0.001 *M* $MgCl_2$, and its absorbancy increases on dialysis against, or alcohol precipitation from, 0.01 *M* EDTA, followed by H_2O (hyperchromic state). Absorbancy also increases gradually on raising the temperature from 0 to 50°. This is in contrast to double stranded nucleic acid, which shows a sharp increase in absorbancy at a definite temperature, usually above 80°, although it is also dependent on salt concentration (Fig. 6-6). Another means of differentiating single and double stranded nucleic acids is by density gradient centrifugation, double stranded nucleic acids showing lower density and sedimentation constants (e.g., 15 S for the double stranded form (RF; see Section XI, B), compared to 26 S for single stranded M12 RNA) (see Fig. 4-2). In the case of RNA, susceptibility to pancreatic ribonuclease (in 0.05 *M* salt) is another useful criterion.

(c) The cyclic nature of a nucleic acid is suggested if it shows sedimentation heterogeneity, with 2 or 3 components, and transition from one to the other on limited nuclease action. The more definitive proof comes from the negative outcome of a search for end groups. Also, double stranded circular molecules are resistant to denaturation, or rather they renature very readily even on fast cooling. Electron microscopy, finally, illustrates the cyclic nature most vividly and facilitates length measurements (see Section VI, D).

(d) A viral nucleic acid may show sedimentation heterogeneity (polydispersity) either intrinsic (see Section VI, D) or attributed to intraviral attack by nucleases, and apparently due to conformational weak spots in the molecule. However, a viral nucleic acid can also appear to be homogeneous and yet represent an aggregate of several molecules. That the physical properties of double stranded nucleic acids are not noticeably altered by a few single chain breaks illustrates the ability of secondary forces to simulate nonexistent molecular homogeneity and integrity. Thus, no conclusion about the molecular weight of nucleic acids, and about their homogeneity in strict chemical terms, is possible until intermolecular hydrogen bonding and salt crosslinking have been ruled out by removal of salts or by heat or dimethylsulfoxide treatment.

D. Specific Structural Features of Nucleic Acids of Various Viruses

1. *RNA Viruses*

For most simple viruses, as exemplified by TMV and the small phages, the molecular weight of the RNA or DNA, as determined by sedimentation methods, corresponds to that expected for a single molecule of nucleic acid per particle as derived from analytical data, such as phosphorus analysis, base analysis, or UV absorbance (147, 138). If end group data are available, they generally support this conclusion. Table 6-1 summarizes the present state of knowledge of terminal ribonucleotide sequences of viral RNA's, some of which were discussed in the preceding section.

A disappointing aspect of these data is the lack of correlation of the known terminal nucleotide sequences with known terminal amino acid sequences in viral proteins. It appears possible that the pppG-G- ends of the RNA of many bacteriophages and the -C-C-C-A ends occurring in both plant viruses and phages serve no information-carrying purpose but play a more physical role. The ends of the STNV RNA appear more unique but also in this case of an unusually small and possibly monocistronic messenger, the first codon does not correspond to the first amino acid of the viral coat protein (AGU = serine, not alanine, GC^{Pu}_{Py}).

Alfalfa mosaic, tobacco rattle, cowpea mosaic, bean pod mottle, and several other viruses are heterogeneous in terms of particle size, and their various components can be separated more or less readily and quantitatively by density gradient centrifugation (see Section VIII, E). It then appears that the smaller particles have shorter RNA chains than the longer. End-group studies have not yet been reported on these various RNA fractions, but the assumption is that each has characteristic terminal nucleotide sequences. The functional significance of such multicomponent virus systems is discussed below.

Another group of small isometric viruses, to which belong bromegrass mosaic (BMV), cowpea chlorotic mottle (CCMV), and cucumber mosaic virus (CMV-Y), are uniform in dimensions but contain three RNA components of different sizes (e.g., 26 S, 22 S, 14 S) (37, 25a). The molecular weights of the two smaller components usually add up to that of the large component (e.g., 1×10^6, 0.7×10^6, and 0.3×10^6), and only the large component is infective. These results strongly imply that the smaller components are fragments of the large one, but it is not clear why and how this specific intraviral fragmentation occurs in these viruses. End-group studies are in progress to further characterize the nature of the fragments.

A possibly related and equally unexplained phenomenon is the susceptibility of Qβ RNA to breaking *in vitro* under the influence of minimal RNase treatment (RNA/enzyme, w/w = 200,000, 0°, 30 min) near specific sites of the molecule, giving at first a 68% piece derived from the 3′-end and a 32% remainder (27, 26). It has been suggested that such selective enzyme sensitivity is due to a particular conformation of the viral RNA. The relatively high sedimentation rate of the phage nucleic acids, and of BMV RNA as compared to TMV RNA (about 28 and 30 S for 1 and 2×10^6 molecular weights), suggests that nucleic acids which tend to give specific fragments have a tighter conformation than TMV RNA.

Turnip yellow mosaic virus (TYMV) is one of the most studied viruses, and yet the status of its RNA is not clear. With considerable care, RNA can be obtained which has a molecular weight of 2.0×10^6, corresponding to the total RNA complement in this isometric 5.7×10^6 dalton particle. However, the sedimentation behavior of this RNA shows a sharp transition on treatment with EDTA or with a trace of ribonuclease, and the data were variously interpreted as suggesting either the opening of a cyclic structure or some unknown collapse of a fixed conformation. (Such a possibility must also be considered a possible explanation for the tripartite nature of the RNA of CMV-Y and the related viruses discussed above.) By the polynucleotide phosphokinase method, close to 1 left terminal unphosphorylated adenosine, followed by cytosine, was found in the 2 million RNA component (456).

When TYMV RNA is degraded into 5 S segments by treatment of the virus with alkali (see Section IV, A) and the RNA is isolated (without EDTA), it is found to sediment mainly as a single fast-sedimenting component, apparently the result of intraviral reaggregation of the 32 alkali-caused fragments into a more compact form than that of intact TYMV RNA (237, 39). This observation, together with those on large molecular influenza and Rous sarcoma RNA (see below), indicates the great danger of uncritical evaluation of sedimentation data.

Although influenza virus, the best known of the myxoviruses, is considerably larger and more complex than these plant viruses, its RNA is similar in amount to that of TMV and might well be expected to represent a single molecule. In fact, it is possible to isolate influenza RNA of 38 S, equivalent to the 1% RNA content of that virus of about 300×10^6 daltons. However, by avoiding or rather removing divalent metals by means of the chelating agent EDTA or by heating, a mixture of 5 or 6 components of varying molecular weight is obtained, and it has now been well established that this multiplicity of RNA species is not an artifact (87, 343). Several biological peculiarities of the influenza virus and its RNA are explained by this finding (see Section VIII, H). In contrast, the RNA of Newcastle disease virus (NDV), a paramyxovirus, occurs as a single homogeneous component of about 6×10^6 daltons, and this RNA remains undegraded by either EDTA or the dimethylsulfoxide (DMSO) treatment to be discussed below.

The RNA tumor viruses, such as Rous sarcoma virus (RSV), contain RNA of approximately 10×10^6 daltons which behaves as a single component under various ionic conditions, including those which dissociate influenza RNA. However, on heating or in the presence of DMSO, a solvent which dissociates base–base interactions, a sharp transition in sedimentation behavior occurs, resulting in a main component of approximately 3×10^6 molecular weight and smaller RNA material (88).

The sedimentation behavior of RSV RNA has been compared with that of other viral RNA's both in glycerol gradients after DMSO or heat treatment and directly in 99% DMSO, a new technique which assures complete denaturation of nucleic acids and thus overcomes the tendency to camouflage breaks and other discontinuities through complexing. Yet, in contrast to Rous RNA, the RNA's of TMV and NDV were unaffected by DMSO. Analysis of RSV RNA by polyacrylamide gel electrophoresis has corroborated these data and conclusions.

These new findings clearly show that the single stranded RNA of viruses can occur either as a single molecule per particle or as multiple molecules possibly partly cyclic and partly open. It is also evident that the multiple components of certain single stranded viral RNA's are interconnected by divalent metals and thus become dissociated by EDTA; and that others are interconnected by what is believed to be short complementary segments and thus become dissociated by conditions or solvents which abolish hydrogen-bonded interactions.

The double stranded RNA of the reoviruses also appears to be present in the virus particle in the form of multiple strands of greatly varying lengths. Three main classes of components have been recognized (three groups of 2.3, three of 1.3, and four of $0.8 \times 10^6 = 14 \times 10^6$ daltons, 1.1, 0.6, 0.35 μ long) (405). However, no such uniformity is seen when comparing the RNA complement of individual particles on electron micrographs (33, 92, 485). Reoviruses also contain one or several single stranded adenylic acid-rich components (88.5% A, 10.5% U, 1.5% C) constituting 20% of the total RNA (404, 405).

2. *DNA Viruses*

The smallest viruses containing DNA are the two groups of bacteriophages containing *single stranded circular* DNA, φX174, and fd. The first evidence for circularity was the resistance of φX174 to exonucleases (110). Only the slower-sedimenting component (10 S rather than 12 S), believed to be linear as a result of one break, is degraded by *E. coli* phosphodiesterase, an enzyme that degrades single stranded DNA from the end. The single strandedness is also indicated by base analysis, since thymine represents about one third of the bases in all of these phages that have been analyzed (φX174, M13, fd). Also, the temperature dependence of the UV absorbance of these DNA's, as well as their sensitivity to chain-breaking agents, both physical and enzymatic, are characteristic of single stranded molecules and resemble typical RNA more than typical DNA (424, 425).*

The best known of the small viruses containing double stranded DNA are the polyoma viruses, with a DNA molecular weight of about 3×10^6. That sedimentation velocity studies indicate the presence of two or three components in the range of 14–21 S cannot be explained alone by the fact that polyoma DNA is circular. However, the additional fact that *double stranded cyclic* molecules usually have a *supercoiled conformation* has supplied an adequate explanation for all the properties of the polyoma DNA and the slightly larger papilloma virus DNA's (5×10^6 daltons) (Fig. 6-7) (506, 486).

The supercoiling of these nucleic acids is due to the presence of fewer turns in the double helix than occur typically in linear DNA and to the tendency of this deficiency to become compensated by a twisting of the molecule in the opposite sense. There are apparently 12 supercoiling turns in polyoma and 20 in papilloma DNA, in proportion to their molecular weight. It is believed that this deficiency in Watson-Crick turns is related to linear DNA existing in two different conformations (A and B), depending on humidity. Supercoiling also occurs in the cyclic forms of many larger cellular nucleic acids. The three components of 20.3 S, 15.8 S, and 14.4 S are now identified as the supercoiled circular, the simple circular, and the linear molecule. A break in one chain (nicked) causes transition from supercoiled to simple circular structure, and another break in the vicinity of the first leads to linear molecules, which would then tend to have cohesive ends (see later).

To further characterize the DNA's of the polyoma-papilloma group, *nearest neighbor frequencies* have been determined for each nucleotide (312, 445). This is done by replicating the viral DNA *in vitro* with DNA polymerase acting on the four nucleoside 5′-triphosphates, each of which in turn carries ^{32}P in the α position (69). The product is then degraded enzymatically into 3′-nucleotides, and the extent of labeling of each nucleotide in each of the four reaction mixtures is a measure of the frequency of the 16 possible doublets (see Fig. 6-8) (230). These studies indicate similarities of polyoma virus to SV40 (simiam virus) as well as to the host cell

* Recent evidence indicates that the small animal DNA viruses of the parvo and adeno-associated virus group (Table 3.1) also contain single stranded DNA of 1.7×10^6 dalton molecular weight (537).

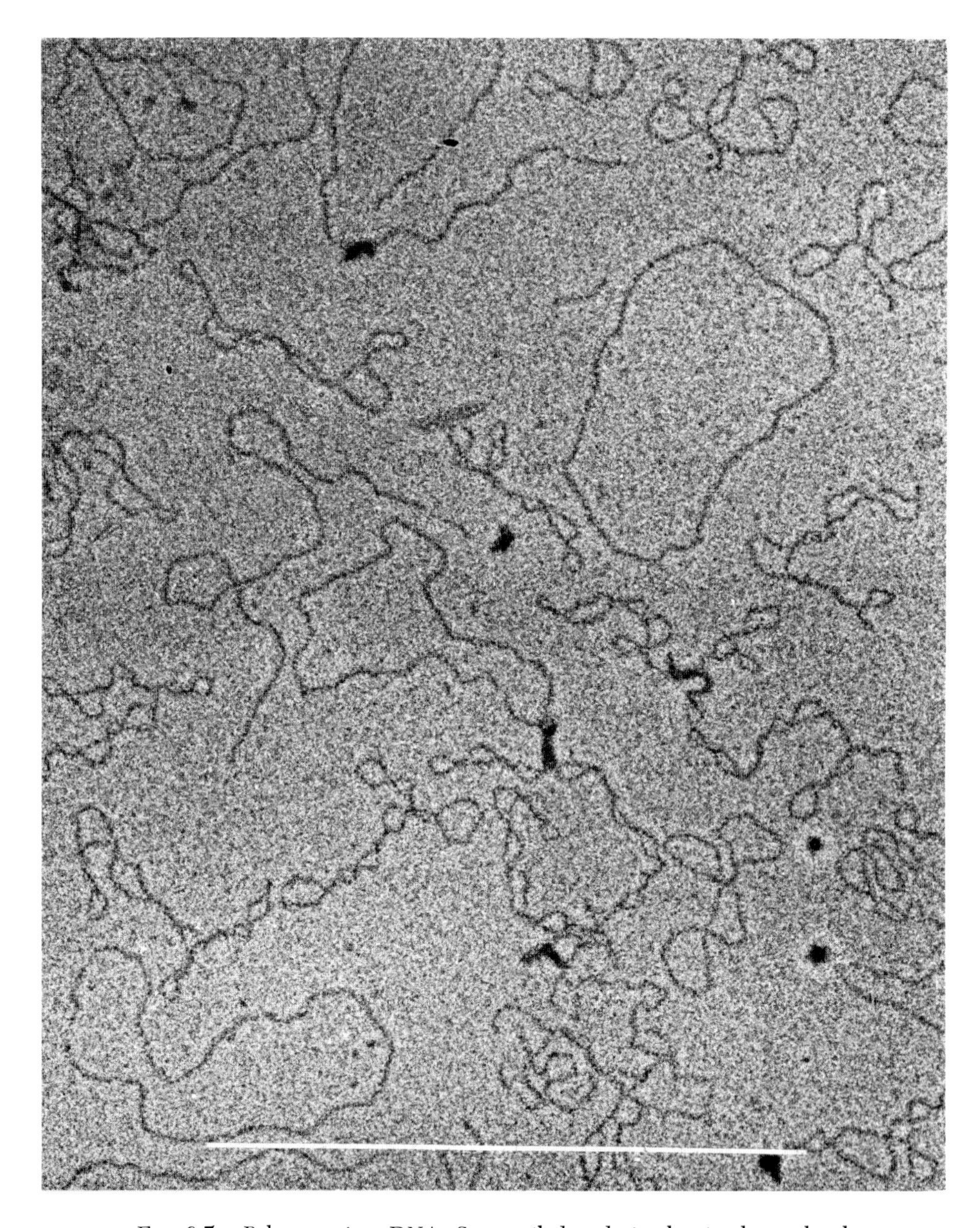

FIG. 6-7. Polyoma virus DNA. Supercoiled and simple circular molecules, as well as a few linear molecules, can be discerned. The bar equals 1 μ. (Courtesy of E.A.C. Follett).

ppp*dA + Template: dG – p-dA – p-dT – p-dT – p-dC – p-dC – p-dT – p-dGp . . .
pppdG + DNA polymerase
pppdT ⟶ ppp – dC – p-dT – p*dA – p*dA – p-dG – p-dG – p*dA – p-dCp . . .
pppdC ↓ + Dnase + spleen diesterase

ppp – dCp, dTp*, dAp*, dAp, dGp, dGp*, dAp, dCp

FIG. 6-8. Nearest neighbor frequency analysis. The four deoxynucleoside triphosphates, with ATP α-^{32}P-labeled, are polymerized on a DNA template and the transfer of label to the 3′ position of the neighboring residues is determined after degradation. Conclusion: In the octanucleotide sequence, A follows T, A, and G; in the template, T precedes A, T, and C. (The asterisk indicates ^{32}P-labeled compounds.)

(human spleen) DNA, and to the pattern given by the papilloma viruses (221).

In contrast, analyses for base compositions and for the tendency to hybridize indicate definite differences between these various DNA's. The *hybridization* studies reveal the existence of similarities of DNA sequences over longer segments than are observed by nearest neighbor frequencies. The principle is to mix two types of denatured DNA (or RNA) and allow them to reanneal. If one of them carries a radioactive label, its incorporation into hybrid double strands can be measured. The double stranded is distinguished from the single stranded material by the usual methods of ultracentrifugation or by nuclease resistance (518, 519).

The adenoviruses show some similarities to the papilloma-polyoma viruses in biological properties, and tend to interact with them, producing mixed particles (see Section XII, B). However, the adenovirus DNA is much larger (about 22×10^6 daltons) and it is not circular. The DNA molecular weights of all three groups corresponds to the DNA content of these groups of viruses.

Many of the polyoma and papilloma viruses and adenoviruses cause occasional transformation of host cells into tumor cells (see Chapter XII). When attempts are made to correlate the various chemical and physical properties with this oncogenic tendency, it appears that the oncogenic adenovirus strains have DNA of molecular weight 21×10^6 daltons and a G + C percentage of 47–49 as compared to 23×10^6 and 56-60% (G + C) for the nononcogenic strains (339). Nearest neighbor and hybridization studies have also been performed; the latter have yielded clear evidence of the close relation of the two oncogenic adenovirus strains (#12, 18), as contrasted to weakly or nononcogenic strains (537).

A yet larger DNA that has been intensely studied, although mainly from the functional rather than the chemical viewpoint, is that of the temperate phage, λ. This double stranded DNA has the expected molecular weight and length for the total DNA of the 57×10^6 dalton particle, namely, 32×10^6 and 17.2 μ (279). λ DNA shows several unusual characteristics. Possibly the most intriguing are the "cohesive sites" (or rather ends) (184), single stranded segments about 20 nucleotides long, at both 5′-ends. These segments are complementary, and they thus give the molecule a tendency to form rings held together by overlapping hydrogen-bonded regions.

Since there is an enzyme, exonuclease III, in *E. coli* which releases nucleotides from the 3′-ends of the double stranded DNA, the appearance of single stranded 5′-terminal segments is not difficult to envisage. However, the previously discussed transition from supercoiled to nicked circular and linear molecules supplies a simpler and more plausible origin for a molecular species of the character of λ DNA, particularly in view of how readily it regains its circular forms (see later).

By use of phosphatase and polynucleotide phosphokinase, the 5′-ends of λ DNA were identified as 5′-phosphorylated A and G, respectively, and the sequence of nucleotides in the cohesive ends has been partly established (526, 527) (see Fig. 6-9). This open

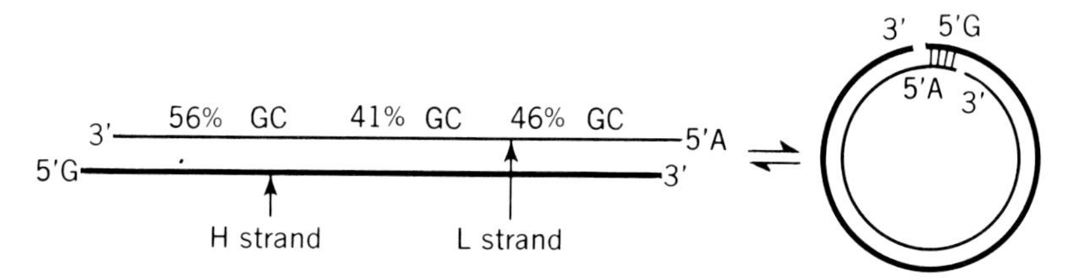

Fig. 6-9. Structural features of the DNA in phage λ. The cohesive ends and the differences in composition, and thus in buoyant densities between the two strands and between the two halves of the molecule are illustrated schematically. H and L stand for the heavy and light strands (184). The author wishes to thank Reinhold Book Corporation for permission to reproduce this diagram.

form of λ DNA occurs only in the phage particle, whereas intracellular phage was found to be circular, through covalent sealing of the cohesive sites (529, 38). Again, the recent recognition that enzymes exist which establish phosphodiester bonds in *E. coli* (328), as well as in eucaryotic cells, make this sealing less startling an event than it would have seemed a few years ago. Also, the tendency of such a molecule to cyclize by the annealing of 20 terminal base pairs is more readily understandable than the intracellular "melting" of this region that transforms the duplex ring with two breaks into the linear form (see Fig. 6-10).

Actually, two circular forms of λ DNA were found to occur intracellularly. The species sedimenting almost twice as fast in sucrose gradients as linear λ DNA was recognized as the twisted supercoiled form (see polyoma DNA), while the species sedimenting 1.14 times as fast was the untwisted circular form, yet covalently closed in one ring since it could not be reconverted into the linear form by heating (38, 326).

Two other properties of λ DNA have played important roles in the recognition of its different physical states and in many other studies with this virus relating to its lysogenic properties. If λ DNA is broken in half by shear forces, the base composition and therefore also the buoyant density of the two halves differs significantly (56% vs 46% G + C) (185, 192). On the other hand, the two complete strands also differ slightly in density (193). This difference can be magnified, and the two strands can be separated in a CsCl density gradient, if a polynucleotide composed of guanylic and inosinic acid residues is added which is bound preferentially by C-rich areas occurring in one of the strands of λ DNA (208, 458). This technique was also found applicable to the separation of the strands of T7 and T4 DNA by using G- and U-rich polymers, respectively (158). The 5′-ends of T7 DNA were shown, by the use of polynucleotide kinase and exonuclease I, to be pApG- and pTpC-, respectively (507).

The DNA's of the large bacteriophages have many singular properties that set them apart from other viral nucleic acids. As previously mentioned, the T-even phages of *E. coli* (T2, T4, T6) contain *hydroxymethylcytosine* in lieu of cytosine. The variously glucosylated state of this base has also been mentioned (T2, T6 25%

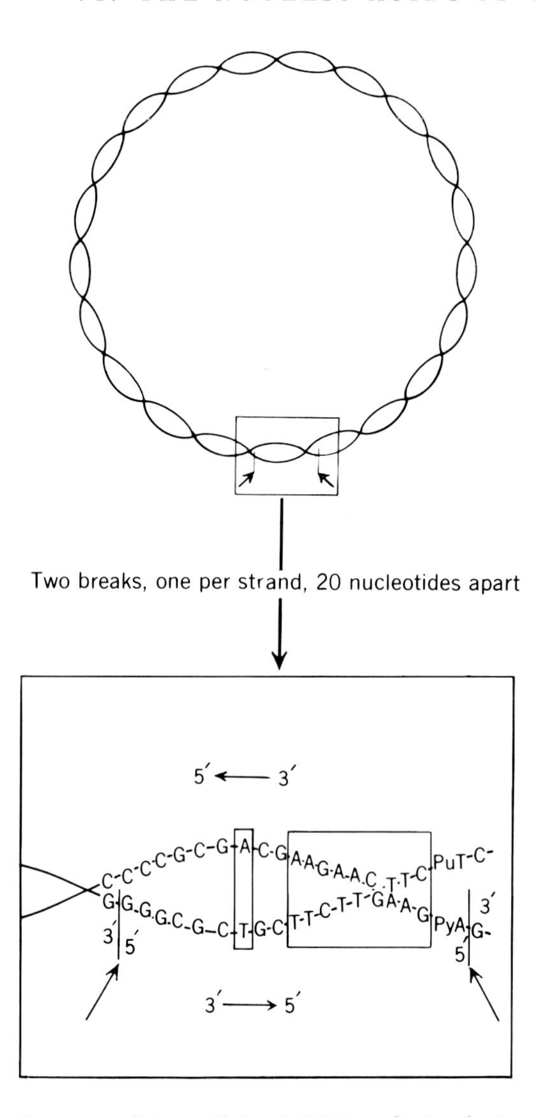

FIG. 6-10. Structure of intracellular λ DNA and of cohesive ends. Intracellular λ is a double stranded circular molecule (largely supercoiled, and not open as shown). In the virus it appears in a linear form resulting from two cuts, as schematically drawn. The sequence of the resulting cohesive ends is presented, as far as it is known, with base pairs of dubious location in boxes (526, 527).

unglucosylated). The composition of T-even phage DNA's is predominantly A + T (about 66%). Small amounts of 6-methyladenine (6-methylaminopurine) are present in T2 and T4 (91).

The determination of the molecular weight of the T-even phage DNA has been a laborious advance in methodology, and has illustrated the dangers of the thin ice of ignorance. Many a paper was written on this subject by some of the most illustrious molecular biologists before it was realized that DNA is degraded by the shear forces developed during pipetting and stirring. Now it is clear that the entire DNA complement of a T-even phage represents one chain molecule of double stranded DNA of a molecular weight of about 120×10^6 daltons and a length of 54 μ (Fig. 6-11) (249, 368, 78). This would seemingly reduce the DNA of this king of the viruses (more like an "organule" than a "molechism") to a pedestrian level of chemical simplicity. But nature has managed to forestall this in magnificent fashion: No two DNA molecules in the T-even phages of a cell need be alike. And yet their disparity makes sense on a molecular level.

The situation is best illustrated by the following example in which the 200,000 base pairs are reduced to 30 symbols: Let the letters of the alphabet (A–Z) stand for a nucleotide sequence. When replicating phage DNA, the cell makes a continuum of possibly 20 alphabets (520 letters). Then this long chain is cut after every thirtieth letter, because 30 letters worth (of nucleotides) fits into one phage head. Thus one phage complement of nucleotides will read A to Z, A to D; and the next, E to Z, A to H; and the next, I to Z, A to L; and the next, M to Z, A to P; and so forth. Each DNA molecule thus starts and ends with a different nucleotide sequence, and each contains the same letters, in order, at both ends. Since the phage specialists have always been exceptionally literate and word-conscious people, two compound terms have been coined to describe this situation: *terminal redundancy* (or repetition) and *circular permutation* (443). The fact is that genetic studies on the T-even phages had strongly indicated that gene A was close to gene Z, and circularity supplied the simplest explanation of these and related data (443). Also, circularity was being demonstrated as a possible or actual structural form of nucleic acids, ranging

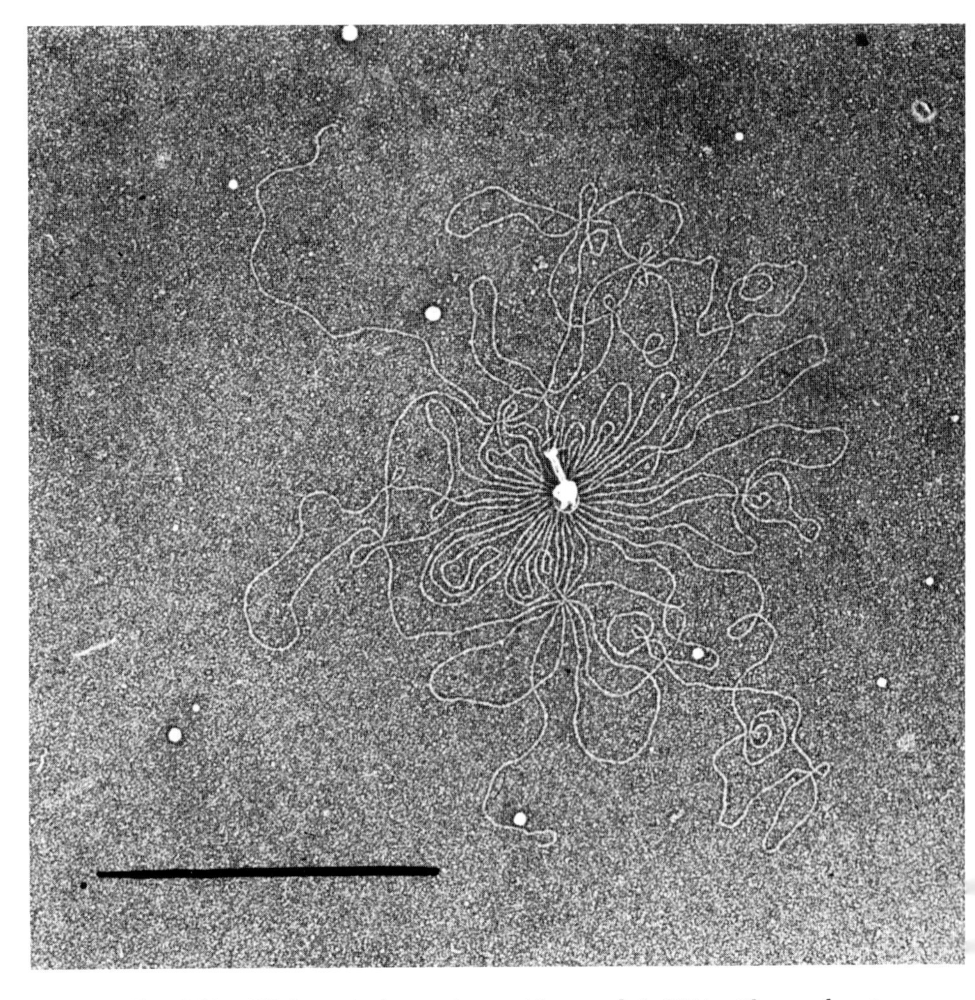

FIG. 6-11. T2 bacteriophage ghost with extruded DNA. The marker is 1 μ in length, the DNA 54 μ (249). Note the ends of the molecule at the top and bottom.

from very small (ϕX174 and fd) to very large ones (the *E. coli* chromosome). Yet, when the T-even phages were osmotically shocked, even in the most gentle manner, the two ends of the extended DNA molecule could always be plainly seen (see Fig. 6-11) (249). There was no doubt that the DNA in the T-even phages was linear, and not circular. This did not preclude the intracellular existence of a circular form, as has been found for λ DNA. However, terminal redundancy could not be explained in that manner and, when evidence for the early appearance of superlarge linear phage DNA was recently uncovered in infected cells (131, 132), the explanation given at the beginning of this section supplied the most plausible, if not the only plausible, description of the state of the DNA of these phages.

The terminal redundancy of the T-even phage DNA's has proved to be a very useful tool in the study of the relationships and structural features of the various phage nucleic acids. As previously mentioned, exonuclease III from *E. coli* degrades each strand of double stranded DNA from the 3′-end, thus creating single stranded 5′-terminal segments. If the degraded segments from both ends of the duplex chain have the same nucleotide sequence because of terminal redundancy, then complementary single stranded segments or cohesive ends are generated which are able to anneal. This reaction can thus be used to test for terminal redundancy. On electron microscopic or sedimentation analysis of such material, many cyclic molecules are detected, in strong support of the existence of terminal redundancy (Fig. 6-12a–c) (280).

Cyclization was also observed with the exonuclease III-treated DNA's of T3 and T7 phages, although here, and particularly at higher concentration, dimers and trimers (concatemers) were also frequently formed, and they occur very rarely with the T-even phages. This result can readily be explained only if all molecules have the same terminal sequences, that is, if they are not circularly permuted. That this conclusion is correct, and that the DNA's of T3 and T7 differ from those of T2 and T4 in showing, like λ DNA, a fixed terminal sequence was also borne out by the results of another type of experiment.

For this purpose the intact (not enzyme-treated) phage DNA molecules were thoroughly denatured with alkali and then reannealed in the gentle manner now preferred over heat for such experiments, namely, by holding the neutralized solutions at 25° in 7.2 M $NaClO_4$. The result was that linear double stranded molecules of equal length predominated for T3 and T7, but the molecules were circular for T2. The noncomplementary single stranded seg-

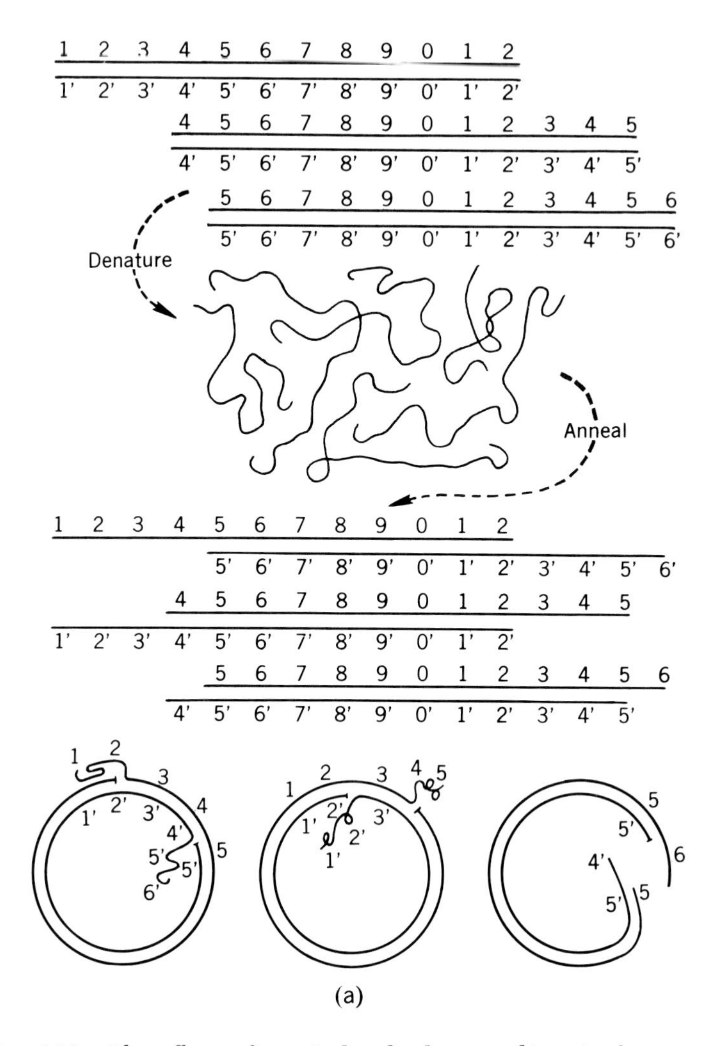

FIG. 6-12. The effects of terminal redundancy and/or circular permutation on denaturation or exonuclease III treatment followed by renaturation of phage DNA (280, 354). (*a*) Circle formation by denaturing and annealing a permuted collection of duplexes. Notice that each permutation is also terminally repetitious. These repetitious terminals cannot find complementary partners and are left out of the circular duplex. Their separation depends on the relative permutation of the partner chains.

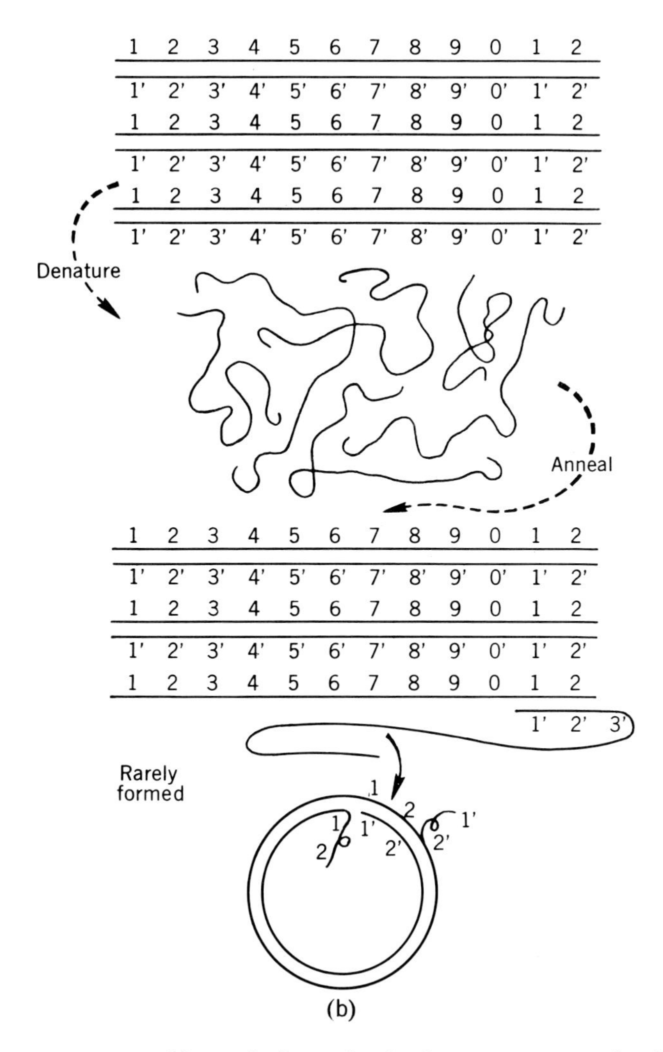

(*b*) Reconstitution of linear duplex molecules from a nonpermuted collection of terminally repetitious single chains. Denaturation and chain separation produce complementary single chains with identical starting points. Notice that the improbable collision between a beginning and an end can (in principle) lead to a circle. The chance of this should depend on the fractional length of the terminal repetition.

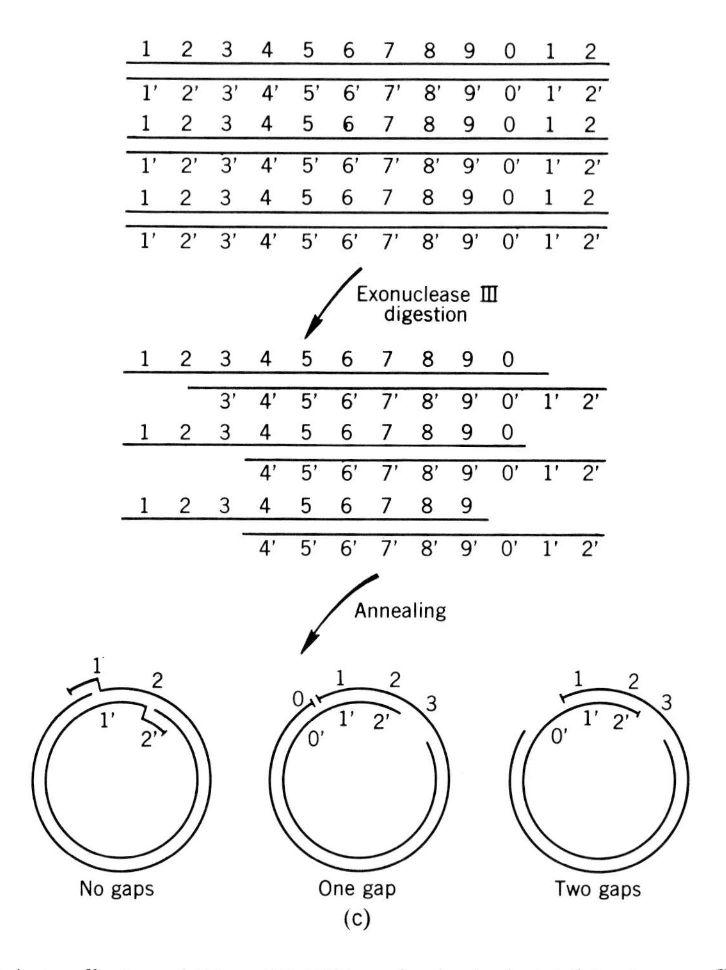

(*c*) A collection of T3 or T7 DNA molecules is depicted by two parallel lines corresponding to the complementary chains of the duplex. Each molecule is shown having the same sequences (nonpermuted), and being terminally repetitious for regions 1 and 2. Exonuclease III exposes the complementary 5′-ended chains at the ends and circle formation takes place on annealing. If the degradation proceeds beyond the limits of the terminal repetition, then two single chain "gaps" will bracket a duplex segment, the length of which is the length of the terminal repetition.

ments resulting from such circularization of T2 appeared as "bushes" on the electron micrographs, with the expected predominant frequency of 2 per circle (see Fig. 6-12). The randomness of the "interbush distances" was regarded as an indication that any, or at least many, parts of the DNA molecule could appear in the repetitious region, as would be expected on the basis of the presumed origin of terminal redundancy discussed above (354). It has been suggested that all phages may be regarded as terminally redundant, the repetitious part at times becoming lost through single strandedness in cohesive ends. Cyclic permutation also is a property of some, but not of all, phages. A diagrammatic presentation allows the correlation of the three parameters, circular permutation (yes or no), terminal redundancy (yes or no), and virulence as contrasted to temperateness, for those phages that have been tested, including

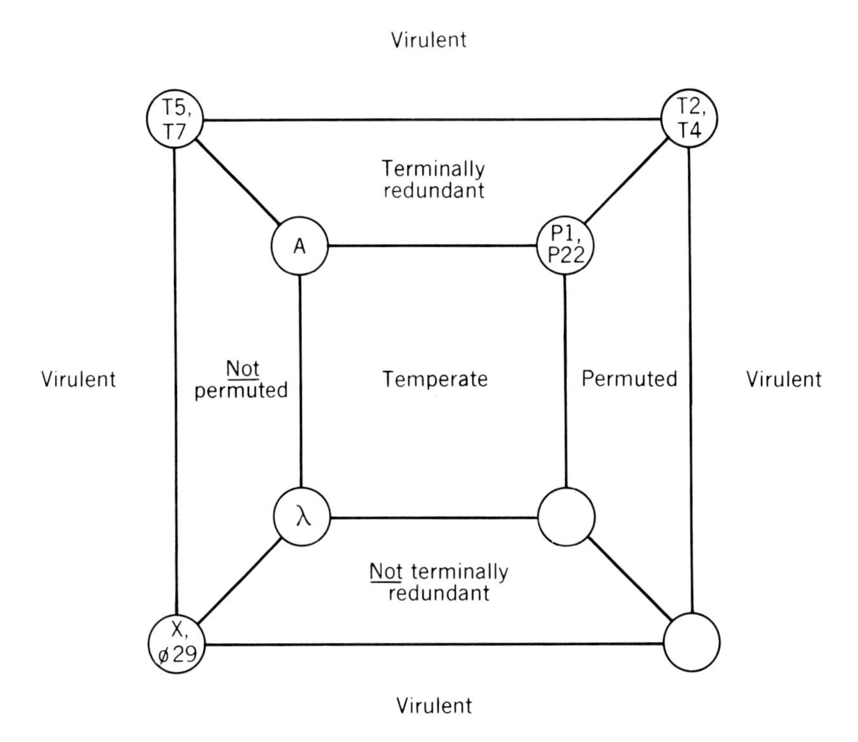

FIG. 6-13. Correlation scheme relating temperate and virulent phages, terminal redundancy, and circular permutation for various phages. (Courtesy of C. A. Thomas.)

a recent study with P22 (352a). Interesting correlations can be derived from this scheme, and they can be added to as new data become available (Fig. 6-13).

The largest DNA molecules would be expected to occur in the largest viruses, the vaccinia and paravaccinia group (3200 and 2150×10^6 daltons), the viruses that were detected by the light microscope many years ago. This is actually the case, since DNA preparations of molecular weights of $175–192 \times 10^6$ have been obtained from these viruses. This means that these largest viruses also contain what appears to be one single molecule of double stranded DNA. Beyond the molecular weight and length (75–80 μ) (31, 214), and analytical data (36% G + C), little is known about this group of linear DNA's (227).

CHAPTER VII

OTHER VIRAL COMPONENTS

Although most plant viruses and most of the RNA phages consist only of one greatly predominant type of each, nucleic acid and protein, many other viruses show a more complex composition. As discussed, the presence of several capsid proteins appears to be characteristic of many of the small animal RNA viruses, the small DNA phages of the ϕX174 type, the adenoviruses, and the reoviruses. Other viruses show an increase in complexity of structure; that is, their nucleocapsids are covered by envelopes which usually consist predominantly of proteins associated with lipids and carbohydrates as well as enzymes.

The most intensively studied of these groups are the myxoviruses and paramyxoviruses, but the RNA tumor viruses, as well as the somewhat smaller encephaloviruses and the much larger DNA viruses of the herpes group, show similar properties. One of the distinguishing properties of several of these virus groups is their sensitivity to ether and low concentrations of detergents, a consequence of the lipid content of their envelopes (206, 207, 269). Other important properties of most of these virus groups are hemagglutination, neuraminidase action, and hemolytic activities, all associated with the viral envelope. Extraction of myxoviruses with hydrocarbons (e.g., heptane, petroleum ether) removes little of the lipids, mainly cholesterol, and does not degrade or inactivate the virus. However, extraction with ether, particularly in the presence of detergents, degrades and inactivates these viruses by completely removing the lipids (about 20% of the weight of the particle). It then becomes possible to separate at least two main components of the influenza virus in the aqueous phase: the hemagglutinin, also carrying the neuraminidase activity (35–65 mμ in diameter; 5% carbohydrate, 95% protein), and the nucleocapsid (363). Yet it has clearly been established that the hemagglutinating and the neuraminidase activities are not properties of the same molecules (400, 86, 262, 501).

The preeminent question in regard to all the viral components discussed in this chapter is whether they are intrinsically virus-specific substances the synthesis of which is elicited by the viral genome, or whether they represent normal cell components which are incorporated more or less accidentally during virus assembly.

The myxoviruses and encephaloviruses, both rich in lipids, are assembled in the lipid-rich layers of the cell near the cell membrane, and are released by budding through that membrane. Thus it is no surprise that the great variety of lipids found in influenza virus (practically all known types of neutral and phospholipids, as well as cholesterol) are similar, quantitatively and qualitatively, to the lipids occurring in the host cell, to the point of reflecting the differences in the composition of different host cells (Table 7.1) (242). Beyond question, much cellular material does become incorporated in the large RNA and DNA viruses which are released by budding from animal cells. However, that does not mean that virus specific glycolipoproteins cannot also exist and play functional roles, and recent studies on Sindbis virus indicate the presence of such a protein as the main secondary component of that virus (besides the nucleocapsid protein) (441). Quite possibly there exists in this case a virus-specific glycoprotein of particular affinity for lipids and capable of picking up whatever lipids it encounters in passing through the cell membrane.

The general conclusions to be drawn are that the approximate composition of the complex viruses in terms of lipids and polysaccharides is largely determined by their genomes and that in some manner this composition represents an evolutionary advantage.

TABLE 7.1
LIPIDS OF INFLUENZA VIRUS AND HOST CELL[a]

Lipid	Chick embryo host		Calf kidney host	
	Host cell	*Virus*	*Host cell*	*Virus*
Fatty acid (% of total)				
Myristic	0.3	0.3	0.7	0.9
Palmitoleic	1	1	1	1
Palmitic	32	38	23	23
Linoleic	9	9	24	27
Oleic	29	29	21	21
Stearic	13	10	21	17
Arachidonic	6	7	6	7
C_{22} polyene (a)	4	5	—	—
(b)	1	1	—	—
Phospholipids (% of total)				
Lecithin	48	28	44	36
Phosphatidylserine	6	7	4	5
Phosphatidylethanolamine	18	22	16	8
Phosphatidylinositol + lysolecithin	6	6	5	5
Sphingomyelin + lyso-phosphatidylethanolamine	20	35	17	14
Phosphatidic acid	—	—	2	12
Unidentified	—	—	9	17

[a] Modified from Ref. 358.

Nevertheless, many or most of the specific lipid and mucoid (glycoprotein) component molecules of these viruses may be typical cellular components. Quite possibly such cellular components as are picked up by the emerging virus during exit from an exhausted or moribund cell may facilitate entry into a healthy cell. The question, how much of the viral substance is new and how much borrowed, loses some of its relevance, if the borrowing is regarded as a genetic trait endowed with biological purpose and advantage.

Considerable work has been done on the lipo- and glycoproteins of the envelopes of influenza and other myxo- and paramyxoviruses, but few definitive conclusions can be made owing to their complexity (for reasons discussed above) and technical difficulties. Also, one strain of a given virus differs markedly from another, the differences varying with the particular host cell type as well. For a thorough description and analysis of this complex field the reader is referred to recent reviews (358, 109).

Neuraminidase is an enzyme which releases *N*-acetylneuraminic acid (sialic acid) (Fig. 7-2) from mucoproteins (152). Its role in virus infection is not yet clear, although it is surely connected with the passage of the virus in and/or out of the cell (75, 76). An attempt to summarize the conclusion to be drawn from very extensive studies on the nature of this enzyme in a few words must state that the neuraminidase is probably virus-specific, because antigenically different neuraminidases were found in different influenza virus strains, as well as in Newcastle disease virus, all grown on the

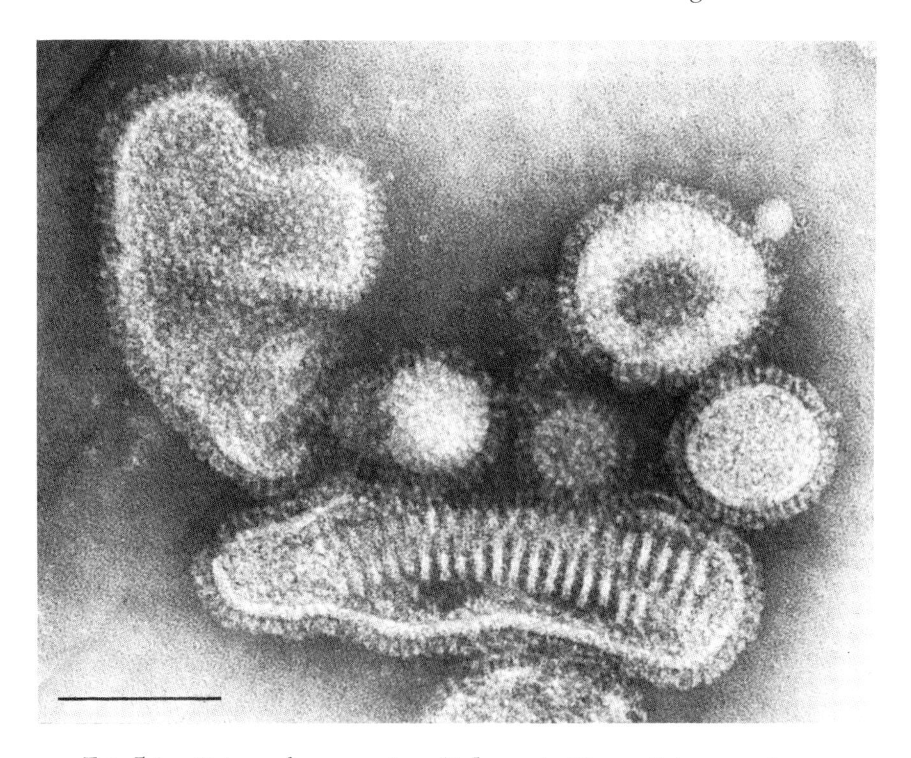

FIG. 7-1. Virions of a myxovirus (influenza). The particles are pleomorphic and the envelope is studded with projections ("spikes"). Penetration of the envelope of one particle by phosphotungstate reveals the coiled internal component (the internal or ribonucleoprotein antigen). (bar = 100 mμ.) (Courtesy of Mrs. J. D. Almeida and F. Fenner.)

same type of chick embryo cells (400, 86, 262, 501). Yet neuraminidase occurs, or can be formed, in the uninfected host cells, as in most animal cells.

The case for the hemagglutinins of the myxoviruses also being virus-specific is at least as good, but less is known about their mode of action. Both types of activity are more or less sensitive to denaturants (including the detergent, sodium dodecyl sulfate, used frequently in degrading the virus). But denaturation can be reversed, and most of the serological specificity and hemagglutinating activity can be recovered from virus dissociated by 67% acetic acid on dilution and neutralization (96). Hemagglutination is a property of many animal viruses, ranging from the smallest to the largest. However, the conditions favoring hemagglutination vary markedly, as illustrated by the fact that influenza virus hemagglutinates at 4°, but the complex is dissociated at 37° (with enzymatic damage to the cell surface by the neuraminidase), while measles virus optimally hemagglutinates at 37°. In line with expectation, this virus contains no neuraminidase. A property frequently associated with viral hemagglutination is hemadsorption (erythrocytes becoming attached to virus-infected cells), a useful tool in animal virology.

The hemolysin activity of the myxoviruses is most probably derived from cellular lysosomes, enzyme-filled vesicles with lipoprotein membranes (319, 320). Another typical cellular component, a mucoprotein of very high carbohydrate content termed the host antigen, has been isolated from influenza strains and characterized in detail (252, 176). All these studies rely extensively on the use of serological methods in identifying and differentiating released proteins. Thus the term antigen appears frequently in lieu of protein, glycoprotein, mucoprotein, or lipoprotein, as the case may be.

Electron microscopy (negative staining) of many of the enveloped viruses (including the myxo-, rhabdo-, and encephaloviruses) reveals regularly spaced "spikes" projecting about as far as the envelope is thick (7–10 mμ) (see Fig. 7-1) (496). It appears probable that these "peplomers," about 2000 per typical influenza virus particle (105), carry the hemagglutinating activity. In contrast, less than a thousand molecules of neuraminidase, located between these spikes, are estimated to occur per particle (502).

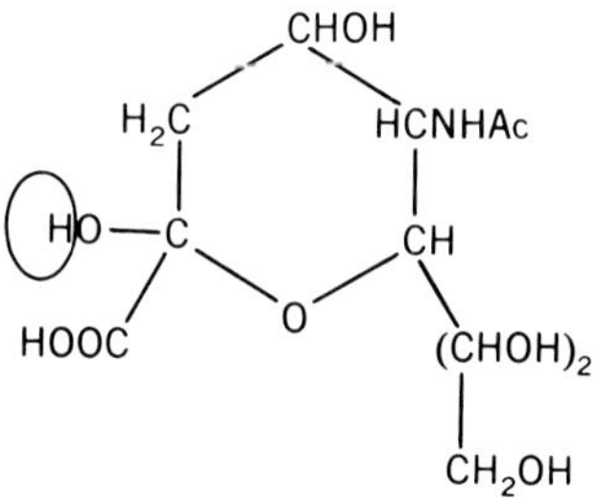

CHOH
H_2C HCNHAc
HO—C CH
HOOC O $(CHOH)_2$
CH_2OH

FIG. 7-2. Neuraminic acid. Neuraminic acid is in glucosidic linkage with the polysaccharide at the site of the circled H atom.

Most of the preceding discussion dealt with the most intensely studied members of the enveloped viruses, the myxoviruses. Both the avian and the murine types of RNA tumor viruses (leukoviruses) seem to contain large amounts of lipid (30–35%, apparently not equal in amount within a population of particles), but little carbohydrate. The protein(s) occurring in the viral envelops are termed to be strain (or type)-specific because they show serologically detectable differences between strains, in contrast to the internal nucleocapsid protein(s), which differentiate only virus groups (group-specific) (89). It appears that the proteins associated with the tumor virus envelopes are predominantly virus-specific, and not derived from the host cell.

The envelope of the few encephaloviruses that have been studied resembles myxovirus envelopes in that it carries hemagglutinin activity, but it appears not to become dissociated by ether, even though these viruses contain much lipid (28% in Sindbis virus, predominantly phospholipids and cholesteral, 3:1). Other indications of the greater stability of viruses of this group have also been reported.

The herpes virus group differs from the previously discussed viruses in these respects (499): the virus particles may, but need not, be enveloped, both types coexisting and being infective; enveloped particles can be detected in the nucleus, in the cytoplasm, or at the cell surface, indicating that these envelopes are not a consequence of maturing and budding at the cell surface. Yet there seems to be little doubt that the components of the herpes virus envelopes are largely or entirely host-derived, and it seems possible that they originate from perinuclear material. In contrast, the envelope components of the largest viruses known, the pox viruses, which contain about 2% each of neutral and phospholipids, seem to originate in the cytoplasm and show little relation to the virus. Various enzymatic activities, such as ADPase and ATPase, are detected in enveloped viruses, but they are generally cellular in nature. The neuraminidases of the myxoviruses and the lysozymes of the phages probably are the only genuinely viral enzymes, except for recent evidence that RNA polymerases are important functional components of the reovirus and probably also the vaccinia virus particle (549).

CHAPTER VIII

THE PROPERTIES OF VIRUS PARTICLES AND THE NATURE OF VIRAL INFECTIVITY

A. Viruses with Helical Nucleocapsids

The helical tube represents the simplest principle of virus construction and the first to be understood. This is largely due to the ready availability of TMV, and to the great advances made since the mid-1940s in the technique and interpretation of the X-ray diffraction analysis of large molecules or small particles (58, 59, 250, 134). The labeling of certain sites of the TMV protein with heavy metals such as lead and mercury has facilitated the task of establishing its internal architecture.

The comparison of the radial electron density distribution of TMV with that of nucleic acid-free TMV capsids has supplied clear evidence for the location of the RNA (Fig. 8-1). As illustrated in Figs. 8-2 to 8-4, the TMV rod has a maximal and minimal diameter of 18 and 15 mμ and consists of 130 turns of a helix in which 49 subunits of coat protein are arranged in a helical repeat unit of 3 turns with a pitch of 2.3 mμ (59, 250). The protein subunits are threaded onto the nucleic acid molecule, a helix with 8 mμ diame-

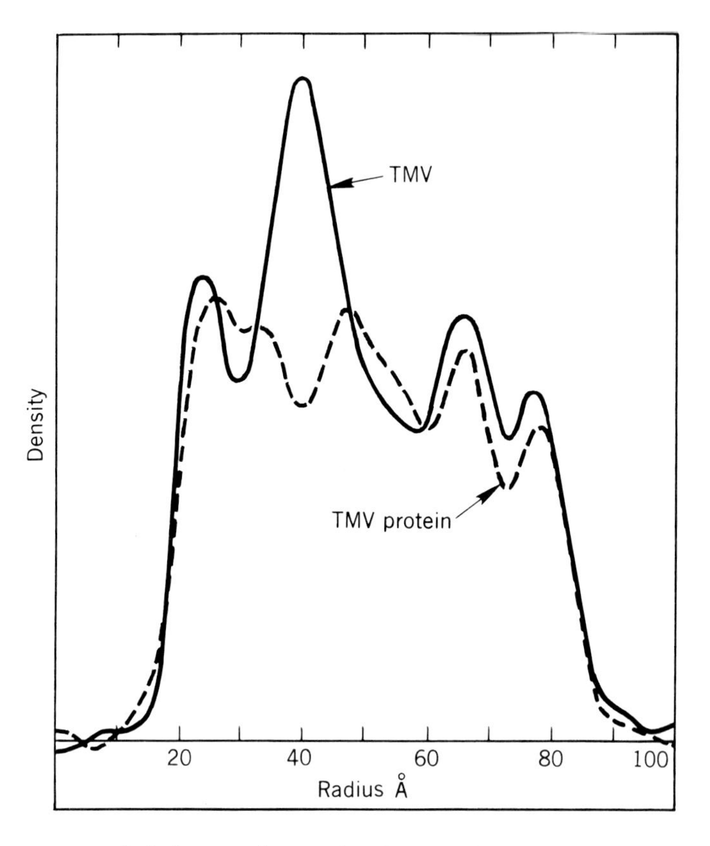

FIG. 8-1. Radial electron density distribution of TMV and TMV protein. Note the absence of a maximum at 4 mμ in the protein rod; it proves that this is the location of the RNA in the complete virus (133).

ter, in such manner that three nucleotides are bound per protein molecule—probably to two arginine and one lysine residue. These data allow calculation of the total length of the virus particle on the basis of a determined molecular weight of 2.05×10^6 for the RNA, and this length agrees with the length of 300 mμ first derived from physical measurements and confirmed in 1946 by electron microscopy. Actually, the best figures now available for TMV indicate the presence of 2130 protein molecules, 6390 nucleotides, and a particle weight of 39.4×10^6 daltons (59, 260). These dimensions

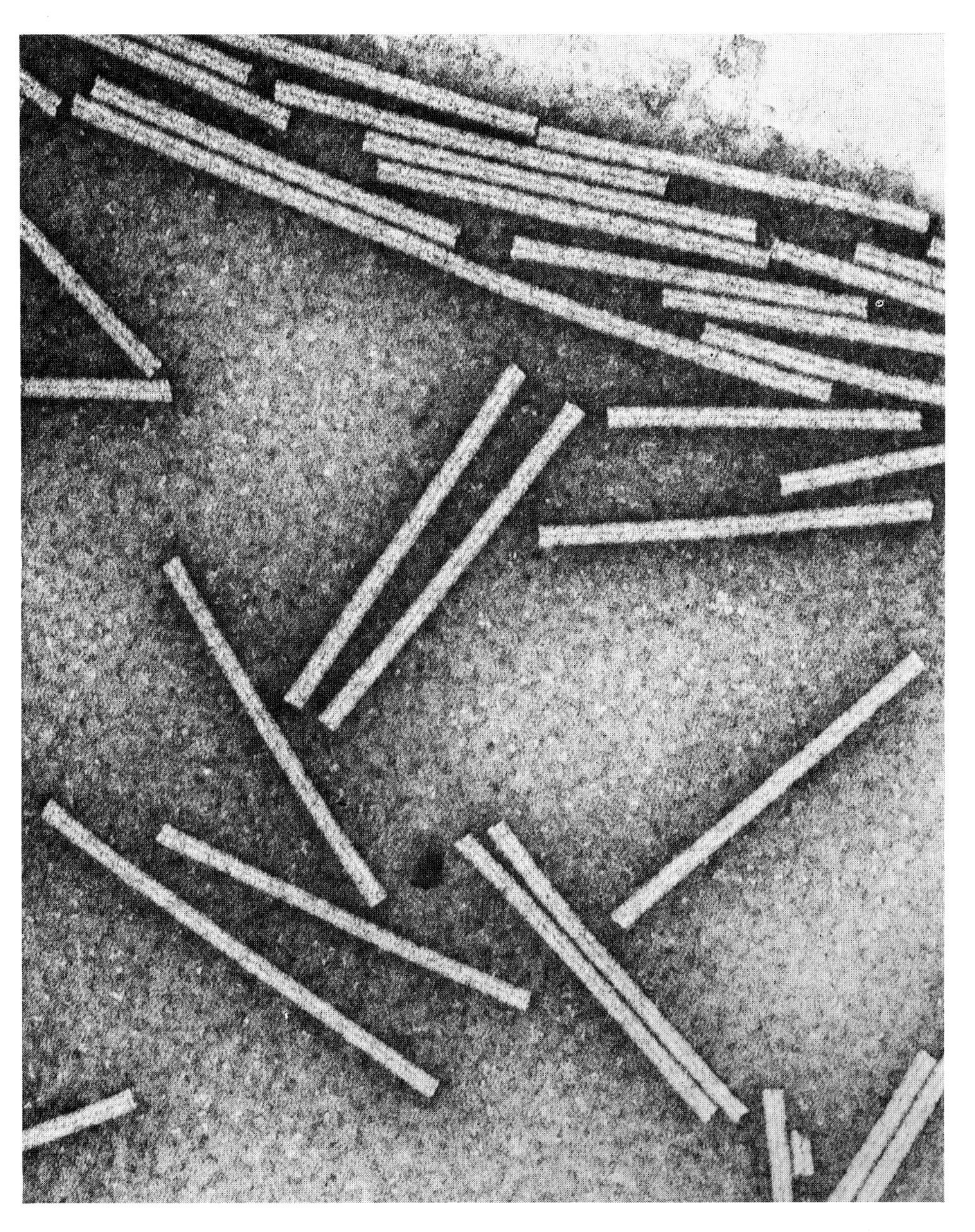

FIG. 8-2. Tobacco mosaic virus rods, negative staining. Note axial empty and thus stain accepting tube, about 4 mμ in diameter according to Fig. 8-1. Typical rod length, 300 mμ.

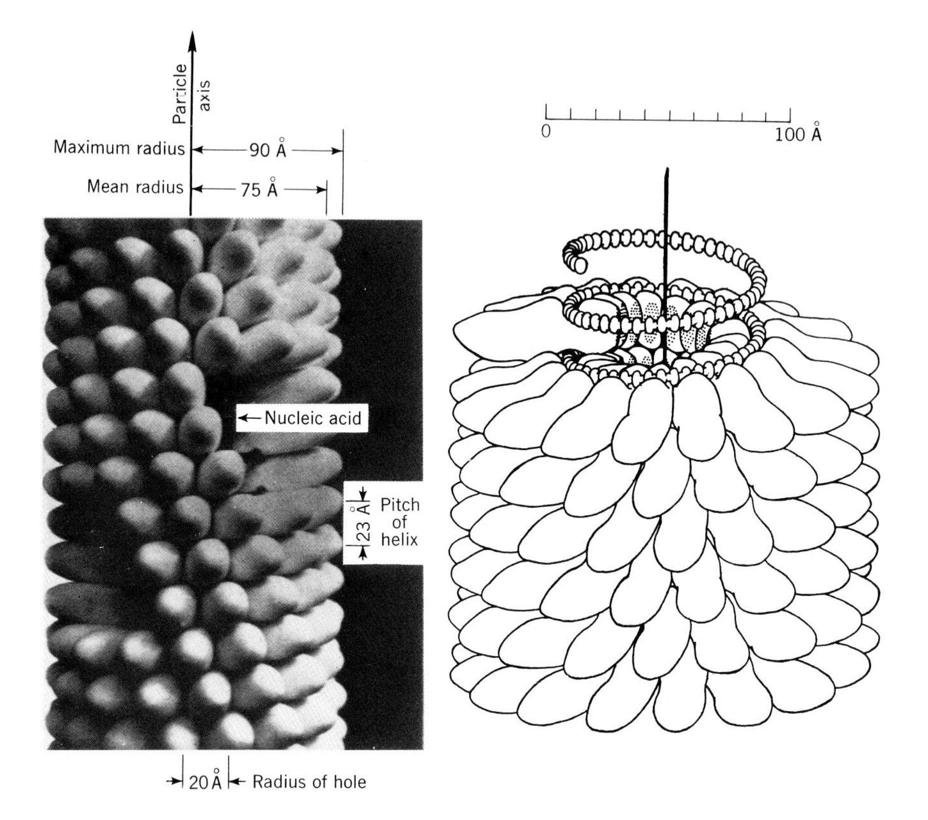

Fig. 8-3. Segment of tobacco mosaic virus particle, model and drawing. Note external protein molecules and internal RNA helix (diameter 4 mμ) (59, 250, 133).

appear to be the same for the many strains of TMV, and for some related viruses not generally classified as strains (133).

Tobacco rattle virus, though unrelated to TMV, has a similar helical arrangement of subunits with a repeat period of 3 turns (76 subunits/3 turns), and a helical pitch of 2.5 mμ. The diameter of the rod is 25.6 mμ (111). The length heterogeneity of this virus has important implications which are discussed later (see Section VIII, E) (325).

Another group of helical viruses, exemplified by potato virus X, appearing as long flexible threadlike rods (11 mμ diameter), has the same pitch as TMV (2.3 mμ), but a repeat unit of 2 turns. The subunit molecular weight in 2 *M* guanidine HCl was reported as 52,000 (349, 350), but more recent studies favor a value of about 23,000 (308).

A seemingly quite different class of viruses, the paramyxoviruses, though consisting predominantly of roughly spherical particles, have nucleocapsids showing the helical pattern. These large viruses (including the Newcastle disease, mumps, measles, and parainfluenza virus) have a complex envelope containing several active proteins, lipids, and carbohydrates (see Chapter VII). However, their nucleocapsid is composed of flexible nucleoprotein rods of the same dimensions as TMV (18 mμ diameter) (202, 203, 512) (Fig. 8-4). The helical nature of the rodlike capsids of the myxoviruses (7–9 mμ diameter) is less well established (See Fig. 7-1).

Another group of rod-shaped viruses with helical components have been termed the rhabdoviruses (rod viruses). The best known example is the vesicular stomatitis virus (VSV), 180 × 70 mμ, which has one flattened end (bullet-shaped) and is covered by a membrane with spikes (418, 316, 205). Short, noninfective particles are also frequently seen. Recent studies with one of many similar plant viruses (wheat striate mosaic) (266) suggest that the bullet shape may be the result of breakage at a particular weak spot of the rod, 260 × 80 mμ in dimensions, the fragments being predominantly 170 and 90 mμ long. The RNA of VSV is part of a very regular nucleocapsid of 5 mμ diameter. The large "fragment" is of about 40 S, and the weight of the total RNA per "unbroken" particle lies between 5 and 7×10^6 daltons. The interpretation of the complex structure of interlocking helices of this virus is illustrated diagrammatically in Fig. 8-5. The rabies virus may also belong in this class (211).

Finally, the rod-shaped DNA phages of fd type (188, 292, 534) should be mentioned here. They show a 3.22 mμ repeat unit and are believed to be composed of two parallel strands of supercoiled protein α-helices.

B. Isometric Viruses

Most viruses appear approximately spherical on electron micrographs and, if their appearance is quite uniform and shows reg-

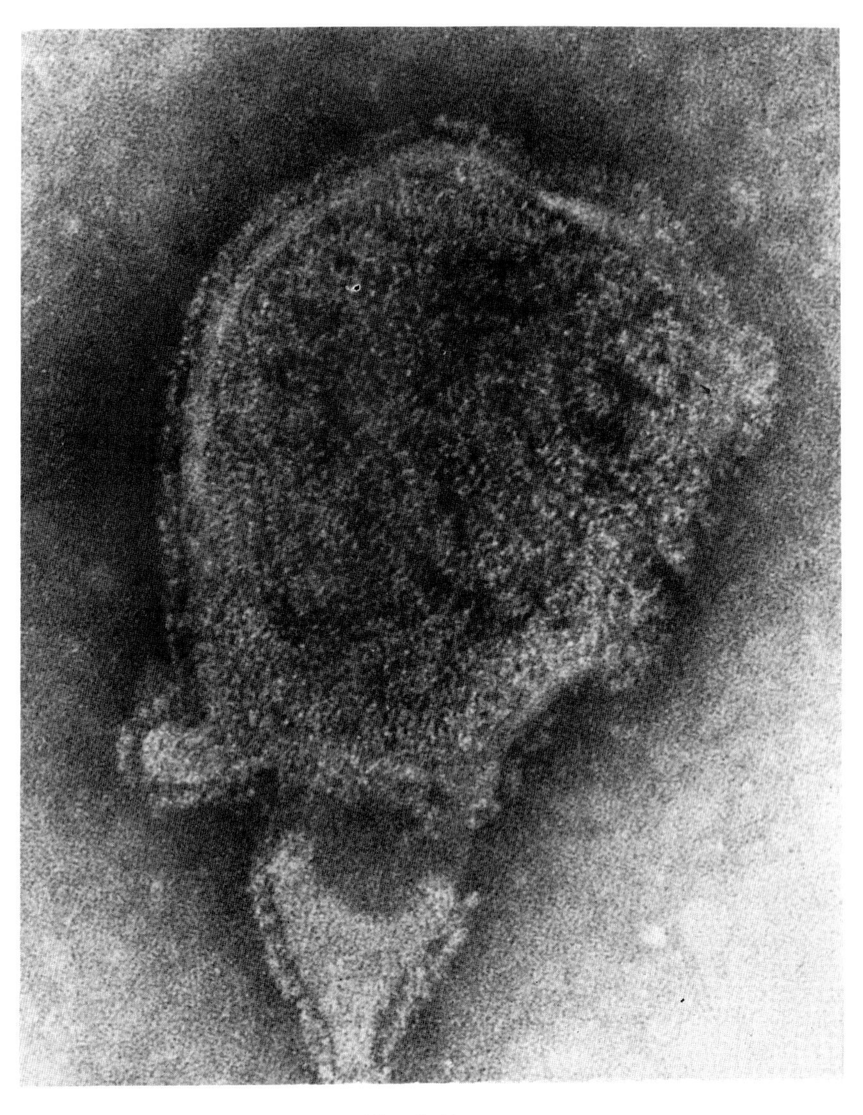

FIG. 8-4A

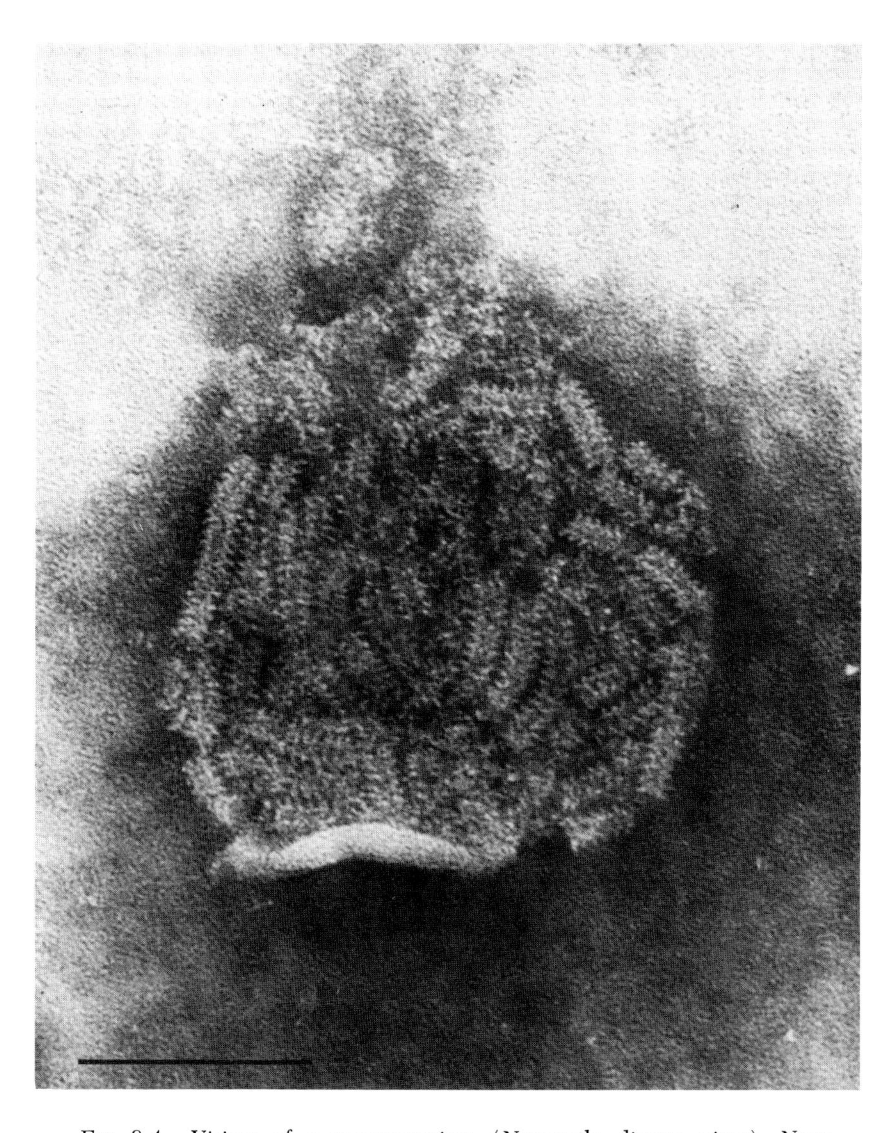

FIG. 8-4. Virions of a paramyxovirus (Newcastle disease virus). Negatively stained. (A) Envelope with peplomers surrounds the internal ribonucleoprotein. (B) Envelope is degraded and the helical arrangement of the subunits of the ribonucleoprotein are visible. Bar = 0.1 μ. (Courtesy of Mrs. J. D. Almeida and F. Fenner.)

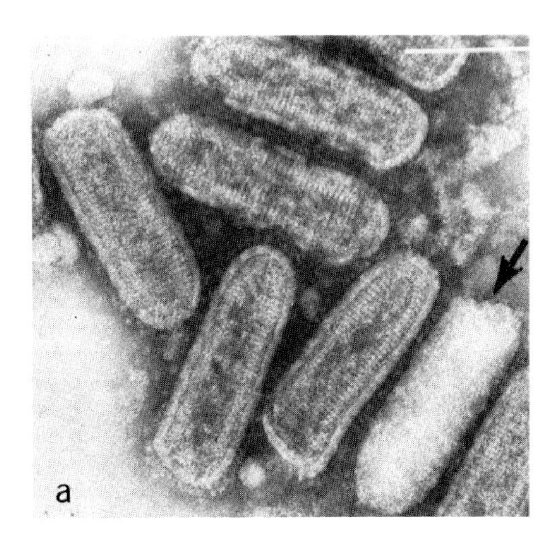

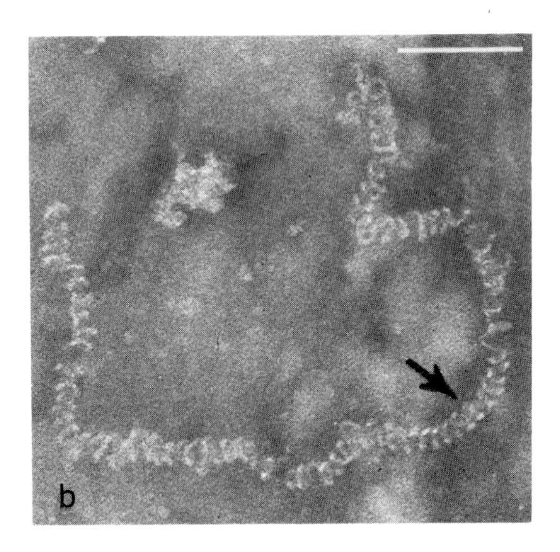

FIG. 8-5. Electron micrograph and diagram of a rhabdovirus. (a) Vesicular stomatitis virus particles prefixed with Regaud's fluid and negatively stained with phosphotungstic acid. All particles except one (arrow) have been penetrated by the negative stain, and they reveal the characteristic cross striations of the internal component. ×135,200. (Photograph kindly supplied by Dr. R. W. Simpson of the Public Health Research Institute of the City of

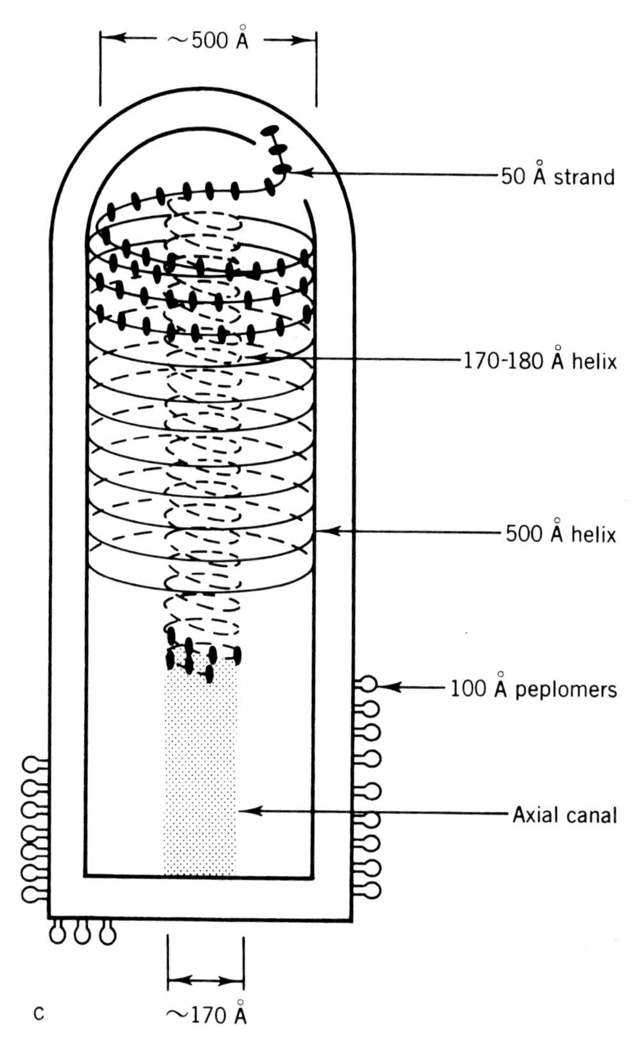

New York.) (b) A free strand of internal component derived from a suspension of VSV prefixed with Regaud's fluid and negatively stained. The arrow indicates a region of the strand where the helical configuration is well preserved. ×135,200. (Photograph from Simpson and Hauser.) (c) Diagram illustrating possible arrangement of nucleocapsid and membranes (109). (Courtesy of F. Fenner.) The diagram illustrates a possible arrangement of the nucleocapsid and outer membranes of VSV. It is assumed that the helices extend the entire length of the virion and that the subunits occur all the way along the 5 mμ strand. The peplomers occur throughout the envelope, which completely encloses the virion. (Courtesy of R. W. Simpson and F. Fenner.)

FIG. 8-6. Tipula iridescent virus and model icosahedron. Both the virus and the model are shadowed from two angles to illustrate the relationship of vertices to edges (516). The diameter of the shadowed particle is about 170 mμ. (Courtesy of R. C. Williams and J. Toby.)

ularly spaced surface features, they can be assumed to be symmetrical polyhedra. The surface features are now recognized as capsomeres, and they usually show readily discernible characteristic differences for different viruses. Generally they are arranged in groups of equilateral triangles, one of the simplest shapes being a dodecahedron, exemplified by ϕX174 (12 capsomeres composed of 15 structural subunits each). Icosa(delta)hedra, bodies with 30 edges, 20 faces, and 12 vertices, represent by far the most common form. It appears that the icosahedral shell, the basic design of the geodesic dome, represents a most efficient design also for biological containers, and one requiring minimum energy in assembly. This was first demonstrated with a large insect virus, the *tipula* iridescent virus (513, 234) (Fig. 8-6).

The quasi-equivalence concept introduced by Caspar and Klug into virology (60) allows the formation of shells with 60 or more building blocks, and this concept accounts best for the geometric arrangement of the capsomeres of most isometric viruses. The basis of this geometry is the coexistence of capsomeres containing 6 subunits on the faces and/or edges of the triangles, and 5 subunits on the vertices. Thus not all capsomeres have identical environments —they are quasi-equivalent. Either the same structural subunits

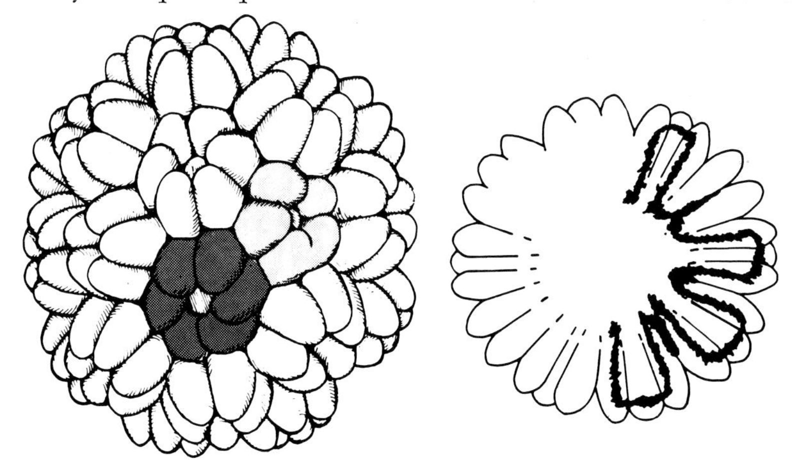

FIG. 8-7. Turnip yellow mosaic virus. Model of the virus with its 32 capsomeres (20 hexamers and 12 pentamers of the 20,000 molecular weight protein chains = structural subunits); and cross section to illustrate the location of the RNA (black strand) (60).

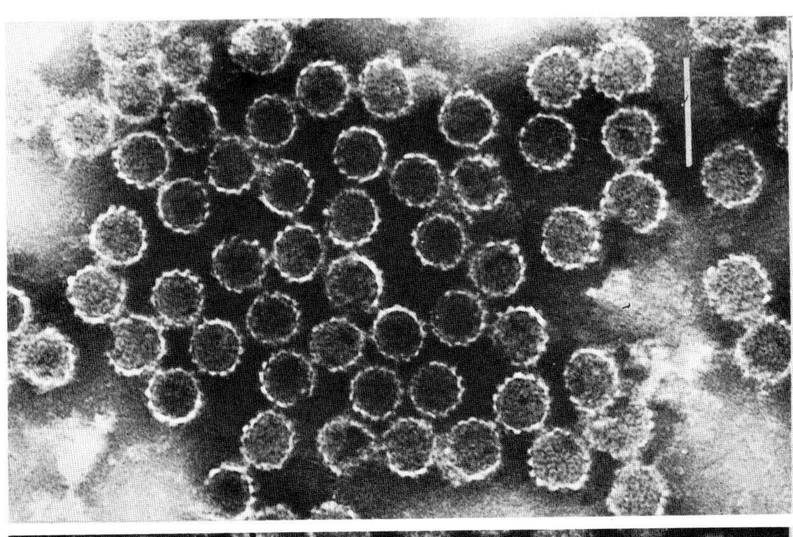

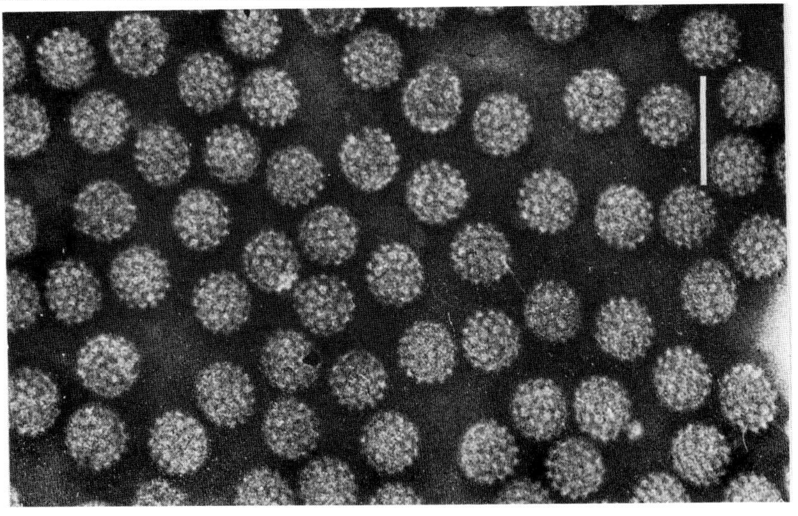

FIG. 8-8. Human papilloma virus particles. The micrograph on the left shows "full" particles, and that on the right "empty," nucleic acid-free, particles. The two preparations negatively stained with potassium phosphotungstate. The bar in each micrograph represents 100 mμ. (Electron micrographs kindly provided by Dr. E. A. C. Follett.)

may make up the pentamer and hexamer subunits, as for TYMV, RNA phages, etc., or different proteins may occur in the two types of capsomeres, as for the adenoviruses (see Section V,B). The triangular faces can be further subdivided into smaller equilateral triangles, the total number, the triangulation number, being *T*. Although the arrangement and number of capsomeres is not equally firmly established for all icosahedral viruses, this number has been proposed as a means of classifying viruses. The principle of icosahedral construction is illustrated by schematic presentations and electron micrographs in Figs. 8-6 to 8-9.

It should be noted that the quasi-equivalency principle allows for some flexibility, and can thus account for the varying shapes among virus particle populations, including oblong bodies and rods, which have been observed to accompany the isometric particles under various conditions of viral infection (187, 40).

C. Viruses Containing More Than One Protein

The preceding two sections dealt with the general architectural principles of typical helical and isometric virus structure. Tables 8.1, 8.2, and 8.3 summarize some of the physical and chemical properties of the simple RNA and DNA viruses. The same principles hold for viruses of increasing complexity, but either more protein shells of capsomeres—the term nucleocapsid loses significance—or envelopes, or tails, and other organelles are added. The structural and, particularly, the important biological aspects of this increasing complexity are discussed in the following sections. The discussion starts with viruses containing more than one type of protein and proceeds to those having more than one molecule of nucleic acid, more than one type of particle, particles with envelopes and various organelles, no particle at all, and, finally, no virus at all.

The simplest viruses containing several proteins are the typical animal picorna viruses, poliomyelitis, mouse encephalitis, etc. The structural or functional role of the four component proteins in these viruses is not clear. Actually, gel electrophoretic analysis of the virus-specific proteins formed intracellularly indicates a much

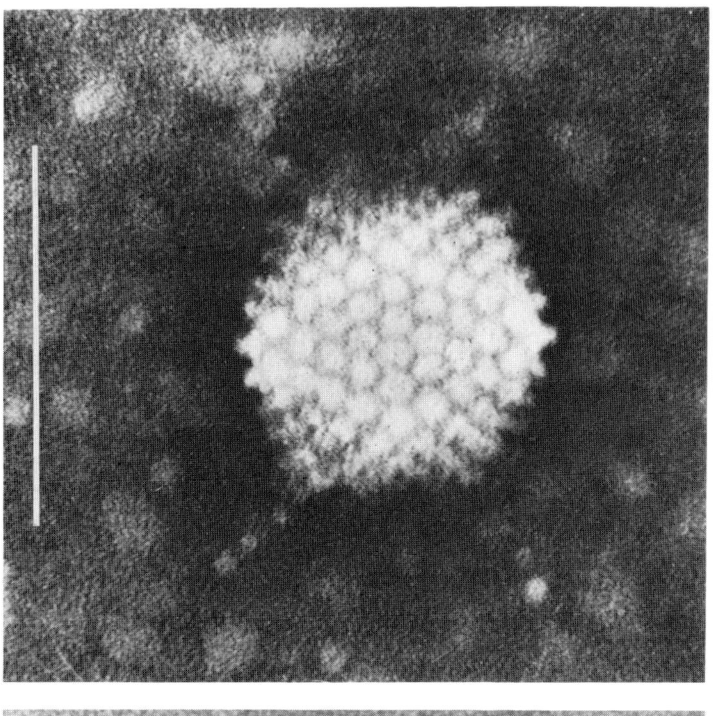

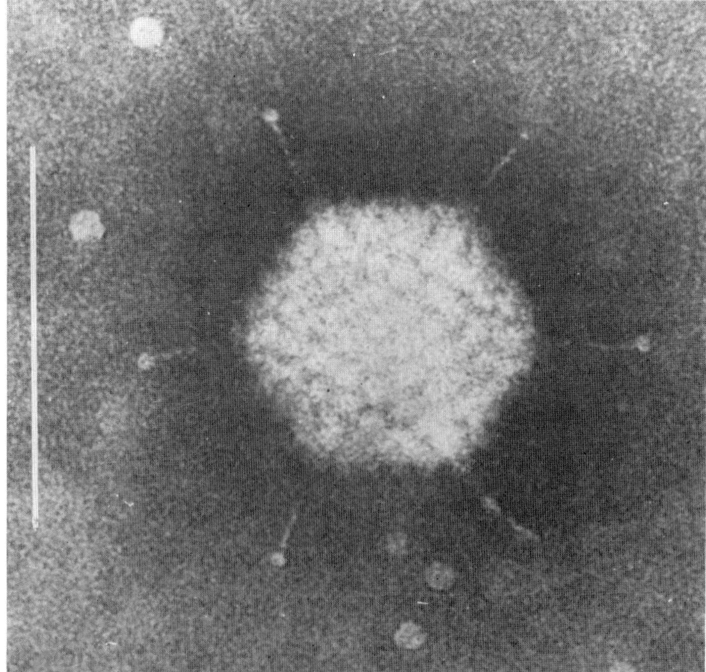

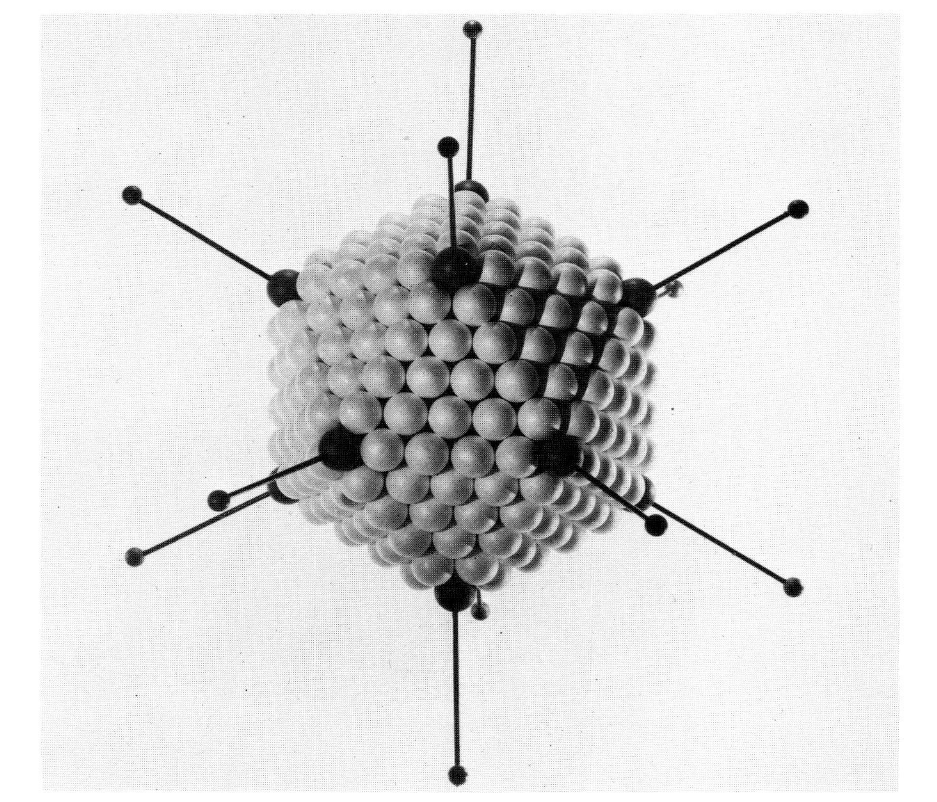

FIG. 8-9. The morphology of the adenovirus particle. The two upper pictures are electron micrographs of adenovirus type 5 particles in preparations negatively stained with sodium silicotungstate. The upper micrograph shows the capsomeres and their icosahedral arrangement, and the lower one the fibers projecting radially from the pentons. The bar in each micrograph represents 100 mμ. The lower photograph is a model of the adenovirus particle based on electron microscopic observations. Kindly provided by Dr. R. C. Valentine (476).

TABLE 8.1
PHYSICAL AND CHEMICAL PROPERTIES OF ONE-COMPONENT PLANT VIRUSES[a]

	Virus particle			RNA						Protein subunit	
						Comp., mole %					
Virus[b]	*Weight* $\times 10^6$ *daltons*	S_{20}	*Diameter, mμ*	*%*	*Mol. wt.* $\times 10^6$	*A*	*G*	*U*	*C*	*No.*	*Mol. wt.* $\times 10^3$
BM	4.6	86	26	21	1.0	—	—	—	—	180	20
BBM	~5	84	26	22	1.0	27	25	29	19	180	21
CM	5–7	~95	28	18	1	—	—	—	—	180	32
TYM	5.6	117	28	34	2.0	21	17	20	42	180	20
						23	17	22	38		
WCM	5.5–7	119	28	35	—	17	16	26	41	180	20
TRS	~5	128	28	34–42	2.1	—	—	—	—	—	19
BS	~8	132	30	16	—	26	28	26	20	—	58
TN	—	118	25–30	—	—	28	25	25	22	—	—
STN	2.0	50	17	20	0.41	28	25	25	22	42	39
SBM	6.6	115	29	21	1.4	24	26	26	24	180	29
TMV	40	190	300 × 18	5	2.0	28	24	28	20	2130	17
Potato X	35	118	540 × 11	(5)	—	32	22	22	24	—	52
Wound tumor	—	—	—	—	—	31	19	31	19	—	—

[a] Rod-shaped (alfalfa mosaic, tobacco rattle) and isometric (cowpea mosaic, bean pod mottle) multicomponent viruses (coviruses) are discussed in the text.

[b] BM = bromegrass mosaic; BBM = broad bean mottle; CM = cucumber mosaic; TYM = turnip yellow mosaic; WCM = wild cucumber mosaic; TRS = tobacco ring spot; BS = bushy stunt; TN = tobacco necrosis; STN = satellite of TN; SBM = southern bean mosaic; TMV = tobacco mosaic.

TABLE 8.2

PHYSICAL AND CHEMICAL PROPERTIES OF TYPICAL ANIMAL AND BACTERIAL RNA VIRUSES

	Virus particle			RNA						Protein subunit	
						Comp., mole %					
Virus[a]	Weight × 10⁶ daltons	S_{20}	Diameter, mμ	%	Mol. wt. × 10⁶	A	G	U	C	No.	Mol. wt. × 10³
FMD	6.1	140	23	32	—	26	24	22	28		(60)
EMC, ME	6.6	160	28	30	2	25	24	27	24	180	26
Polio	6.9	157	28	29	2.0	29	25	25	22	180	27, 32
Sindbis	~50	~300	~45	6	2–3	30	26	20	25		
Reo	80	630	70	(20)	m[b]	28	22	28	22	540	
VS	—	~45	60 × 180	—	5–7	27	20	31	22		
Influenza	~300	—	80–100	0.8	m[b]	23	20	33	24		
RS	(~500)	—	~80	2	m[b]	25	28	22	24		
ND	(~700)	—	100–200	(~1)	6	24	24	29	23		
f2 phage	4.0	80	25	32	1.05	23	26	26	25	180	15
Qβ phage	4	84	25	~28	0.9	22	24	29	25		

[a] FMD = foot and mouth disease; EMC = encephalomyocarditis; ME = mouse encephalitis; VS = vesicular stomatitis; RS = Rous sarcoma; ND = Newcastle disease.

[b] m stands for multiple strands, 3×10^6 or less in molecular weight, amounting to about 11, 15, and 3×10^6 for RS, reo, and influenza viruses, respectively.

TABLE 8.3
PHYSICAL AND CHEMICAL PROPERTIES OF TYPICAL DNA VIRUSES

	Virus particle			*DNA*						
						Composition				
Virus	*Weight* $\times 10^6$ *daltons*	S_{20}	*Diameter, mμ*[a]	*%*	*Mol. wt.* $\times 10^6$	*A*	*G*	*U*	*C*	*Comments*[b]
φX174 phage	6.2	114	25	25.5	1.6–1.7	0.25	0.24	0.33	0.18	s.s. cyclic
fd phage	11.3	40	800	12.2	1.4	0.24	0.20	0.34	0.22	s.s. cyclic
Polyoma, SV40	40		45	8.5	3.4	0.26	0.24	0.26	0.24	d.s.
Papilloma	41		53	12	5.0	0.26	0.24	0.26	0.24	d.s.
Adeno type 2	177		70	13	23	0.22	0.28	0.22	0.28	d.s.
λ phage	57	416	50	58	33	0.25	0.25	0.25	0.25	d.s. coh. ends
T2 phage	215	700	95	56	120	0.33	0.18	0.33	0.17[c]	d.s. term. red.
Equine abortion	~1000		(150)	9	92	0.21	0.29	0.21	0.29	d.s.
Tipula iridescent	~1100[d]		(160)	19	120	0.18	0.32	0.18	0.32	d.s.
Cowpox	~3200		(300)	5	160	0.32	0.18	0.32	0.18	d.s.

[a] Largest dimension for nonisometric viruses.
[b] s.s. = single stranded; d.s. = double stranded; coh = cohesive; term. red. = terminal redundancy.
[c] Hydroxymethylcytosine, largely glucosylated, rather than cytosine.
[d] About 50% water of hydration[234].

greater number, namely 14, of which several are quite large and in too great a quantity to be enzymes (455, 283). The sum of these proteins would require more amino acid sequence information than is carried on the 2×10^6 dalton RNA of these viruses (equivalent to about 2000 amino acids). It is therefore assumed, and in one instance has been demonstrated, that the large components are precursor proteins which yield the typical capsid protein by proteolytic cleavage at specific sites (see Section VIII,K) (222, 223, 284).

The capsids of the RNA phages consist of only one protein, but the virions are infective only when one molecule of the A protein (maturation factor) is present in the particle (see Section X,C). Thus, strictly speaking, these viruses belong also in the group of multiprotein viruses.

Another virus group containing several proteins (at least 9) are the adenoviruses which have quite a complex virion structure, but carry in 23×10^6 daltons of DNA enough information to account for such structural complexity. Little beyond their strong basic nature and their approximate molecular weights and partial compositions is known about the three proteins making up the nucleocapsid, proper, of the adenoviruses (377), and about three other lower molecular weight components of the virion. This is partly due to the esthetic and quantitative predominance of the outer shell of those viruses, with its three striking and different protein components: the 240 hexons, each with 6 neighbors; 12 pentons, at the vertices, each with 5 neighboring hexons; and the fibers, with little knobs, attached to pentons (see Fig. 8-9). Yet there remain doubts also about the molecular weights of these components (476, 285, 286).

Another group of multiprotein viruses are the reoviruses. They consist of a double layer of capsomeres. The more stable inner capsid (45 mμ diameter) consists of 2 proteins and 10 nucleic acids, and it can be obtained in infectious though unstable form by heat treatment (296). The outer shell seems to consist of 92 hollow prisms secured by a matrix of protein which is susceptible to chymotrypsin (see Fig. 8-10). Partial digestion of this matrix activates the virus, possibly by facilitating its uncoating (297, 314, 229), releasing its RNA polymerase (549).

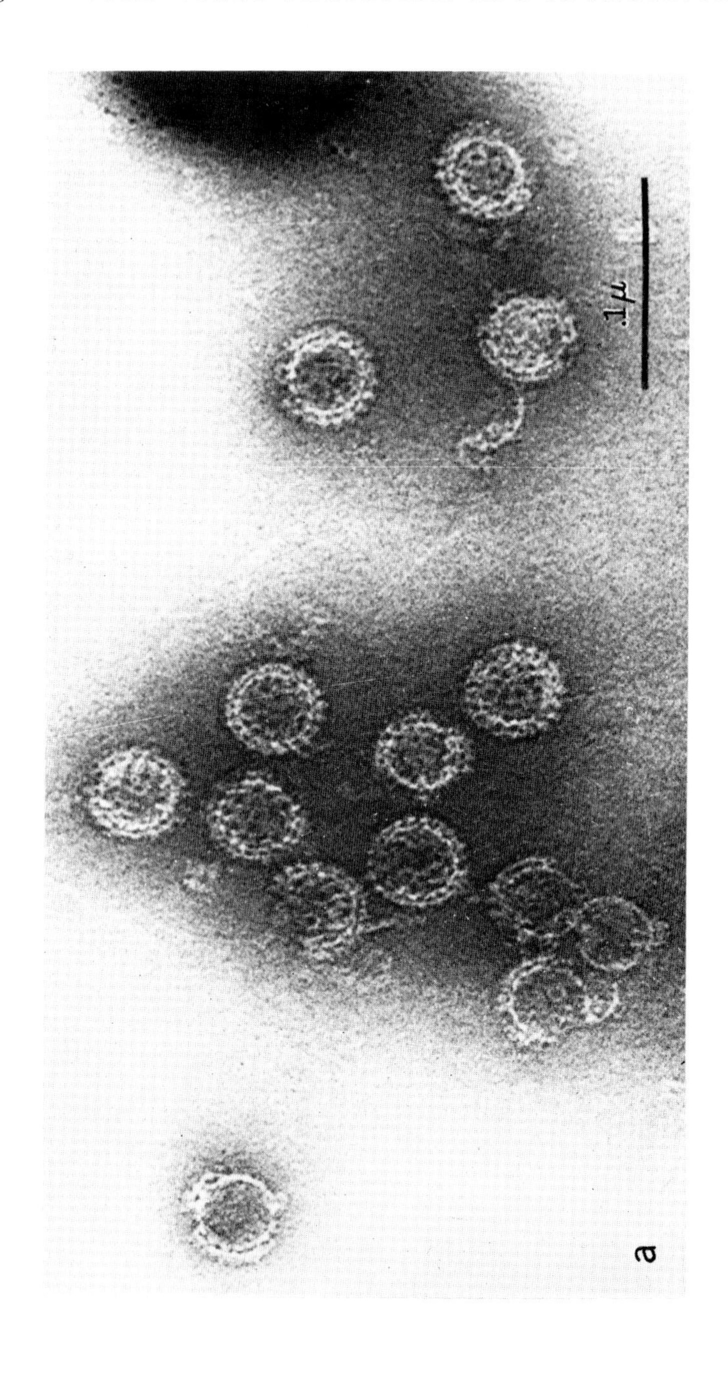
1μ
a

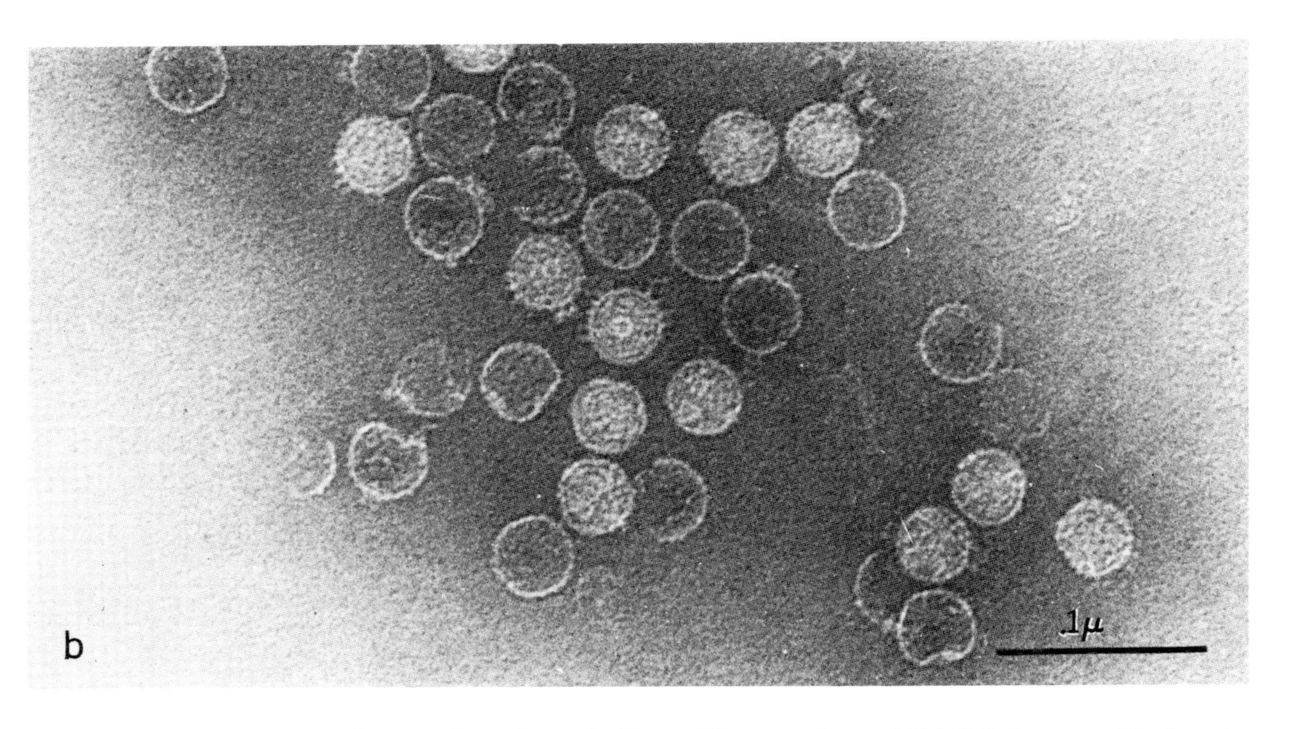

FIG 8-10. Reovirus. a: particles stained to enhance double capsid construction. × 120,000. b: inner capsids of reovirus artificially produced by treatment with trypsin. The particles are 45 mμ in diameter and there is some evidence of subunit structure. Preparation is negatively stained with 2% ammonium molybdate. × 120,000 (297).

All the more complex viruses distinguished by an outer envelope (myxo-, paramyxo-, leuko-, herpes, encephalo-) or by various organelles (pox membranes and phage tails) also contain many different proteins. These viruses are discussed in later sections.

D. Viruses Containing More Than One Molecule of Nucleic Acid

Many viruses consist of several nucleic acid molecules, often covered by only one protein, and these viruses fall into a few distinct classes. One class includes the salt-sensitive plant viruses (example, BMV) which contain RNA of approximately 1.0, 0.7, and 0.3 daltons (37). Only the large RNA is infective; the others are believed to result from intraviral fragmentation, the specificity and the causes of which are not understood (see Section VIII,D).

Another group contains viruses the RNA of which is believed to be genuinely multiple in nature. Among them are the reoviruses and related plant viruses, the influenza viruses, and the Rous sarcoma viruses (see Section VI,D). The RNA of the influenza virus can be isolated as one piece in the presence of divalent metals, but upon treatment with EDTA 5 or 6 components can be resolved (87, 343). These components, isolated as a complex or a mixture, are not infective. It is believed that they all carry viral information and that infectivity results only from a complete set entering a host cell, an event which can easily be assured only by infecting with the intact virus. Several unusual properties of influenza virus, and particularly the high level of recombination observed with strains of this virus, as contrasted to all other RNA viruses, is well explained by these multiple RNA components (see later).

Rous sarcoma virus has also recently been shown to contain several RNA components of up to 3×10^6 molecular weight, the 70 S RNA previously isolated representing an aggregate. In this case, EDTA does not dissociate the complex, but heat or dimethylsulfoxide achieves this, and it is therefore concluded that the RNA components of this and related tumor viruses are held together by the hydrogen bonding of certain complementary segments of the molecules (see Section VI,D) (88). The RNA of RSV, as a complex or dissociated, is also not infective.

E. Viruses Consisting of More Than One Type of Particle, Coviruses

Several groups of plant viruses appear to be of particular interest because they consist of more than one type of particle. One group is exemplified by the cowpea mosaic virus (51, 480), which shows two components of 95 S and 115 S containing 23 and 32% RNA, respectively. Each of these RNA's is homogeneous, and they differ in base composition. Neither virus nor RNA component singly is infective; only the combination of the two viruses or RNA's is. The interpretation of these data is that the viral genome is distributed over the two RNA molecules and that one crucial function, probably the synthesis of the replicase, requires information which is carried partly on the shorter and partly on the longer RNA molecule. The situation seems the same with the bean pod mottle and with alfalfa mosaic viruses.

The alfalfa mosaic virus, AMV, consists of oblong but apparently nonhelical particles of 18 mμ diameter and 36, 42, and 54 mμ lengths. Three and four types of particles are discernible on electron micrographs (146) or sucrose gradients (245), and two components (called bottom and top a) have been obtained in pure form and have been shown to contain different RNA's in amounts approximately proportional to particle size (484). The bottom component showed very little, and the top component a, no infectivity, but combinations of both were highly infective. When the two components were derived from different strains, the progeny showed properties characteristic of each parent. When, after many transfers during which this hybrid virus maintained the same characteristics, it was again degraded, and bottom and top a components were isolated and crossed back, the original strains were recovered. In this study, evidence was also presented by the virus concentration-to-lesion ration (see Fig. 8-11) that multiple AMV particles (more than 1) are required to produce a lesion. It is again concluded that genome components from both particles are required to produce an infection. The strain mixture experiments further show that the coat protein is coded for by the top component a. This conclusion is substantiated by studies of the messenger activity of this component in the *E. coli* cell-free system (see Section XI,B).

The properties of the rod-shaped tobacco rattle virus (TRV) shows an interesting difference from that observed with the cowpea and alfalfa mosaic type of viruses. In the tobacco rattle virus, rods of two different lengths (e.g., 180 and 75 mμ for the PRN strain) contain one RNA component each. When these components, of the virus or RNA, are tested for infectivity, it appears that the short piece is noninfectious, while the long one produces viruslike lesions from which, however, no typical virus, but only an unstable, yet phenol-resistant, infective agent can be isolated. In contrast, the combination of the two components produces typical viruslike and

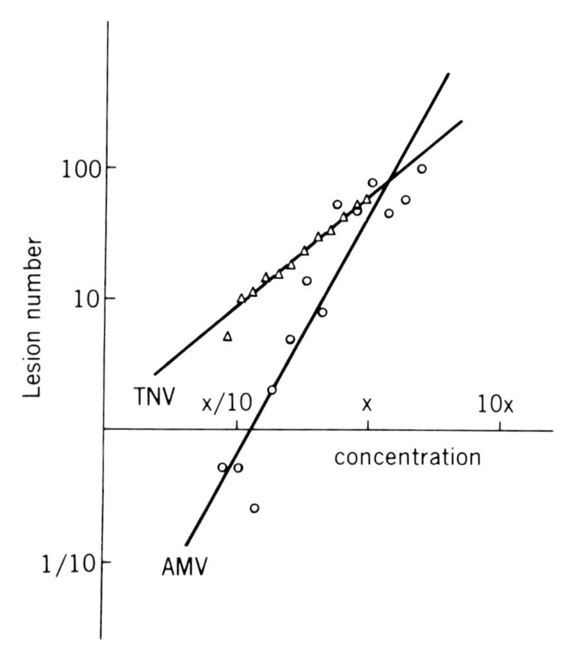

FIG. 8-11. Relation of virus concentration to lesion numbers. Response to inoculation of various concentrations of a covirus system (AMV), as compared to a typical single-component virus (TNV) (484).

virus-containing lesions (271, 140, 381, 544). The preferred interpretation of these findings, which is supported by strain-mixing experiments of the type discussed above for AMV (381), is that the RNA in the larger particle codes for the RNA replicase but lacks

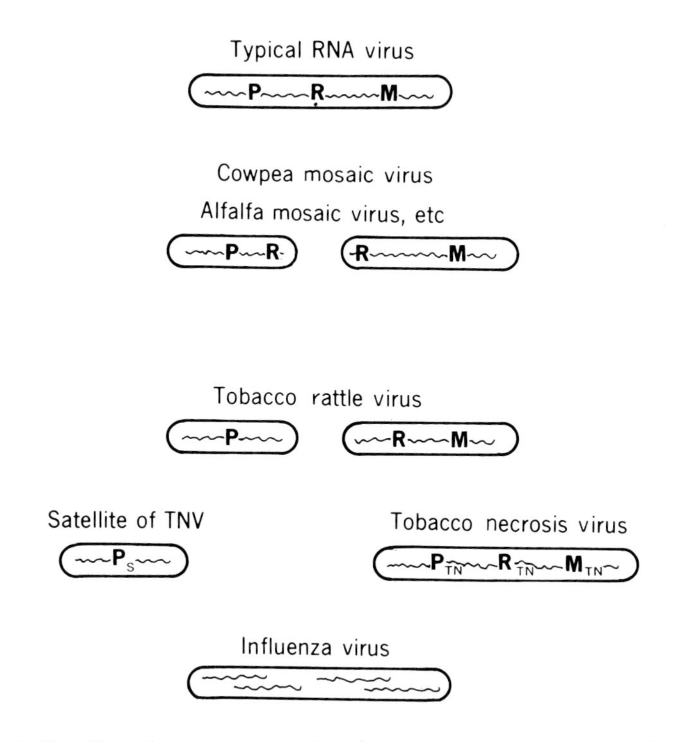

FIG. 8-12. Hypothetical genome distribution in various covirus and satellite systems. P, R, and M represent the three cistrons (genes) known to occur in RNA phages. P, the protein coat gene, and R, the replicase gene, almost surely occur also in plant and animal viruses; M, the maturation factor, may stand for any other genes in general. The scheme illustrates coviruses with the R gene distributed over two types of particles, or completely in one type separate from the P gene. Finally, the satellite which contains only its own P_S, as contrasted to the P_{TN} of the complete tobacco necrosis helper virus. In influenza and other viruses the genome is distributed over many RNA molecules in the same particle.

the cistron for the coat protein, while the shorter RNA codes for that protein. Thus complete virus formation requires both types of particles, but an unstable infection yielding only viral RNA results from infection with the larger particle (or its RNA) alone. Another interpretation would be to postulate that a maturation factor be coded by the shorter RNA. In any case, all plausible explanations suggest that different and definite parts of the viral genome are carried by the different RNA molecules, encased in separate virions. With regard to semantics, a new term would seem to be required for the viruses that require two or more different particles to initiate infection. *Coviruses* is suggested. Thus covirus pairs have been found to carry the required genome for all diseases discussed in this section. The interpretation of the various covirus and satellite systems discussed in the preceding sections and the following one, in terms of genome content, is illustrated in Fig. 8-12.

F. Virus Parasites

The distribution of viral information over several particles is remindful of the nature of the tobacco necrosis satellite virus (STNV). This small viruslike particle (2×10^6 daltons) largely consists of only one cistron (2.7 of 4.0×10^5 daltons) coding for its coat protein (351, 367). This cistron has become entirely dependent on another complete virus, the tobacco necrosis virus (TNV), to supply all other functions of which, presumably, the RNA replicase is the most important. And yet there exists no evidence for any relationship, in terms of coat proteins, between these two particles, the parasite satellite and its supporting virus (241). In a consideration of the limits in the biological capabilities of a simple virus, and its complete dependence on the metabolism of a living cell, it is quite astonishing to observe the existence of a subvirus, a parasite of a parasite, which does not even carry all the genetic information necessary for its own replication. Yet it appears probable that the occurrence of the satellite of TNV is not a singular one. A similar parasitic relationship seems to exist in a group of animal viruses, the adeno-associated viruses (AAV), containing according to recent studies, single stranded DNA of molecular weight 1.7×10^6, which can replicate only in the presence of replicating adenoviruses (427, 191, 190). Here, also, no serological relationship exists between the parasite and its helper. In both cases the parasite represses the replication of the helper virus.

These situations are quite different from the relation of a defective virus such as Rous sarcoma virus to its helper virus, where the failure to code for a (functional) protein is overcome by the production of the helper's protein (see Section XII,C) (365). Also, the role of the helper in permitting infection of *E. coli* by λ phage DNA is different in nature (see Section VIII,14). In none of these cases does the nucleic acid fail in producing its replicase, as it does in STNV.

The terms *deficient* and *defective* could be defined in a manner to cover these possibilities: Viruses which lack the information to make functional replicase (such as STNV) could be termed deficient, while viruses which only fail to elicit the synthesis of one or several structural protein(s) could be called defective. These various forms have recently been reviewed, although the two terms were then used in less clearly discriminating manner (241).

G. Naturally Nonencapsidated Infective Nucleic Acids

Another need to stretch the virus concept has become evident with the recognition that there probably exist cytopathological states which result in, and are transmitted by, free nucleic acid. Owing to the great sensitivity of nucleic acids to various agents, particularly the ubiquitous nucleases, these infective agents are not easy to study. Definitive evidence to validate the belief that a certain disease is caused by an infective nucleic acid rather than by a particularly labile virus is often lacking. The best supported case may be that of the potato spindle tuber virus (81).

This infective agent which sediments very slowly (about 10 S compared to STNV's 50 S) is not harmed by alcohol precipitation or phenol treatment. It is inactivated by RNase only at low ionic strength. On the basis of such data, it is suggested that the "virus" may be double stranded RNA (81, 538). On the other hand, some data obtained with tobacco necrosis virus (241) suggest that certain unstable forms of this virus may represent instances in which single stranded RNA is the natural infective agent. It also seems possible that scrapie, an economically important disease of sheep, is carried by a free nucleic acid. To account for its heat and enzyme resistance, a circular, supercoiled, double stranded RNA might be postulated.

Operationally, the study of these diseases carried by nucleic acids is similar to that of defective mutants of viruses which are unable to produce functional coat proteins. Such mutants have been obtained from TMV and RNA phages (see Sections VIII,K, IX,B).

H. Decapsidated Infective Nucleic Acids

1. *Qualitative Observations*

The first experimental indications that the infectivity of viruses resided in their nucleic acids were supplied by the experiments of Hershey and Chase concerning the mechanism of T2 infection (183). The first step in these experiments was to allow phage which had been grown in the presence of either ^{32}P or ^{35}S, and thus contained radioactive phosphorus or sulfur, to attach itself to the host cell. The infected bacteria were then centrifuged off and subjected to agitation for various time periods in a blender, a procedure which detaches the phages from the bacteria. It appeared that about 75% of the ^{32}P had then entered the *E. coli* cells, but only about 20% of the ^{35}S. Since almost all of the phosphorus of a phage is in its DNA, and most of the sulfur is in the coat proteins, these experiments were interpreted as evidence that on infection the DNA of the phage entered the host cell without the bulk of the viral protein. Later it became clear that minor protein and peptide components of the phage, including the sulfur containing so-called internal protein were actually injected together with the DNA. In the meantime, however, the genetic competence of the DNA had become generally accepted, and now it appears certain that the phage proteins play no decisive part in the intracellular replication of the phage.

The next important advance resulted from studies with a small RNA virus, TMV. This virus can be degraded by comparatively simple methods into its two components. We now know that full length RNA can be obtained from TMV in good yields by any method that dissociates the protein from the RNA at or near neutrality and under conditions which prevent or minimize its degradation by nucleases (115, 147, 129). Enzymes which degrade RNA are frequently associated with viruses, and may also be introduced from the hands or from equipment or reagents used by the operator; bentonite proved to be the most effective means of removing or inhibiting these enzymes when isolating viral RNA and working with it.

When TMV RNA was prepared by methods that did not harm it, and was applied to susceptible leaves, lesions resulted which were indistinguishable from typical virus lesions, and from which infectious virus could again be isolated. Thus the infectivity of an RNA virus was shown to be due to its RNA. When purified viral nucleic acid, which under optimal conditions contained less than 1 protein subunit per 20 RNA molecules (see later), was used for these studies, absolute proof was available for the first time that the infectivity of a virus resided entirely in its nucleic acid. Infective RNA has since been obtained from most plant viruses that have been degraded by similar methods, although the purity of the RNA has not generally been established by such rigorous methods.

The RNA of many of the small animal picorna viruses was isolated in similar manner and found to be infectious. Typical examples are the poliomyelitis, mouse encephalitis, encephalomyocarditis, and foot and mouth disease viruses, all, like TMV, containing RNA of about 2×10^6 daltons molecular weight and showing a sedimentation constant of about 30 S (380, 298). Infectious RNA (30–45 S) has also been isolated from several members of the larger and more complex group of encephaloviruses, e.g., the eastern and western equine encephalitis viruses (503).

Of the DNA viruses attacking animals, some of the parvo-,

polyoma, and papilloma viruses have also yielded infective DNA (1.7 to 5×10^6 daltons) (69, 220).

The cases in which infectivity was not detected in isolated nucleic acids may be divided into two main groups. RNA and DNA molecules of 6–200 $\times 10^6$ daltons molecular weight, exemplified by the Newcastle disease virus (6×10^6 RNA), the adenoviruses (22×10^6 DNA), or the vaccinia virus (180×10^6 DNA), may be unable to enter a cell intact if not aided and protected by the viral coat or envelope. The lesser amount of RNA of some other viruses is now known to consist of several molecules, and the probability of introducing a complete set of genetic components into a host cell without the benefit of packaging can be regarded as minimal. This may be the explanation for the noninfectivity of influenza virus RNA (total about 3×10^6, possibly 6 components), Rous sarcoma RNA (3 components of about 3×10^6), and reovirus RNA (total 14×10^6, 10 double stranded components).

When considering bacterial viruses, it appears that many of the bacteria, including the most commonly used host, *E. coli*, cannot be infected by viral RNA or DNA.* This is probably due to the rigid nature of the cell wall of gram negative bacteria. However, spheroplasts and protoplasts (spherical particles deprived of part or all of their cell wall by lysozyme + EDTA treatment of gram negative or gram positive bacteria, respectively, and maintained in isotonic sucrose solution to keep them from bursting) are susceptible to both the RNA and the DNA of the small phages (the f2 groups, Qβ, φX174, fd, etc.) (425). Spheroplasts are even susceptible to the medium large DNA (about 30–32 $\times 10^6$ daltons) of the phages T1 and λ (530, 48). This is in contrast to the resistance of animal cells to infection by any nucleic acid larger than 5×10^6 daltons, and may be attributed to the cell membrane of the spheroplasts being less complex than the cell wall of mammalian cells. In the case of λ DNA, only cyclized molecules or molecules with cohesive ends are infective, since both removal of the single stranded segment and "repair" synthesis (using *E. coli* DNA polymerase to add nucleotides to the shorter piece, complementary to the single stranded segment) cause loss of infectivity (530).

* In contrast, *B. subtilis* has been reported infectable by the DNA (25×10^6 daltons) of a newly discovered phage (546a).

One unexplained feature of the infectivity of λ DNA is its ability to infect intact *E. coli* in the presence of a helper phage, the action of which can be genetically differentiated from that of the DNA donor. Only the linear form is highly infectious in this assay, and then only if its cohesive ends are the same as those of the helper phage DNA (84).

2. *Quantitation*

So far the infectivity of viral nucleic acids has been discussed only in terms of yes or no, without reference to quantitative aspects. However, such a qualitative approach is always dangerous in science, because it favors errors and misinterpretations. Although influenza virus RNA was described above as noninfectious, many papers can be found in the literature on the isolation of infective 'flu RNA. The key to this puzzle is probably one of quantitation. Thus occasionally a molecular complex of 'flu RNA components may escape dissociation, whether by incomplete removal of the capsid protein or for other reasons, and such a complex may be able to enter a host cell intact and start an infection. It is therefore important in this type of work to measure and express infectivity in quantitative terms, and to define clearly the experimental conditions used. Furthermore, whenever the infectivity of a nucleic acid preparation is low, a number of tests can and should be applied to verify that the properties of the infective agent are truly those of a nucleic acid, and not those of the virus from which it was prepared, because the possibility always exists that a few virus particles remain undegraded and that such particles are responsible for the infectivity of the nucleic acid preparation.

Ideally, each molecule of a viral nucleic acid should initiate an infection, whether localized or systemic. However, as previously stated, the highest absolute efficiency attained with intact virus particles was slightly less than 50% (about two T2 phage particles being required to produce a plaque) (276), and in most cases it is 1–0.001%. It might be surmised that evolution had generally favored protein-coated virus particles because of their greater efficacy, and on this basis the isolated nucleic acids would be expected to be less efficient than the corresponding intact particle.

This has generally been the case, about 100–1000 times as many nucleic acid molecules being required to initiate infection than are required of the intact virus (147, 115). This ratio, however, is strongly dependent on the test conditions. One of the major reasons for the inefficiency of viral RNA is the presence of nucleases in all tissues, and test conditions which depress or inhibit nuclease action favor the relative infectivity of viral RNA. The polyacidic clay, bentonite, has been found similarly effective for the purpose of depressing nucleases during assay, as it does during RNA preparation (419). Another means of increasing the relative efficiency of the RNA is to use host cells of comparatively low RNase content. Finally, assay media can be selected which are selectively unfavorable for the virus (such as high pH) and thus the RNA seems relatively more efficient, but this actually amounts to cheating because the absolute efficiency of the RNA is not increased by such procedures.

In general, it can be stated that the more stable the virus and the more effective its protein coat, the lower is the relative infectivity of its naked nucleic acids. Thus BMV and other members of the group of plant viruses that are dissociated by neutral salts are not appreciably more infective than their isolated RNA's.

There also exists among the TMV strains an interesting example of the role of the protein in affecting this ratio, namely, the Holmes ribgrass strain (HR). The infectivity of the nucleic acids isolated from wild type TMV and from HR is the same. However, comparative assays of the intact viruses show HR to be only about one twentieth as infective as common TMV, a fact which is attributed to a less firmly packed protein coat (see Section V,B1 concerning the differences in amino acid sequence of these two coat proteins). Thus the relative efficiency of HR RNA as compared to HR virus is twentyfold greater than that of TMV RNA (compared to TMV), even though their absolute efficiencies are the same.

The conclusions which can be derived from this discussion are that the infectivity to be expected for a viral nucleic acid cannot be predicted beyond the point that it may be 1–4 orders of magnitude less than that of the intact virus. If it is quite low, then the demonstration that infection is actually due to nucleic acid molecules becomes particularly important. The methods used for this purpose are now discussed.

3. *Proof for Infectivity in Viral RNA or DNA*

The most decisive single test for the nucleic acid nature of an infectious agent is its sensitivity to nucleases or phosphodiesterases. The nucleic acids inside the virion are more or less completely protected from extrinsic nucleases. Once conditions of enzyme treatment have been established which do not decrease the infectivity of a given virus, the loss of infectivity by the nucleic acid preparation obtained from that virus on enzyme treatment under the same conditions is tantamount to proof that the infectivity was not due to residual intact virus particles (147, 130a).

Another useful test to establish the nature of viral infectivity employs immunological methods. When viruses are administered to an animal, e.g., a rabbit, their coat proteins evoke the formation of antibodies associated with the gamma globulin fraction of its serum. The interaction of these antibodies with added virus leads to inactivation of the virus, and to precipitation of the complex at sufficiently high concentrations. However, viral RNA or DNA is not affected by the antiviral antibody. Such tests are complicated by the fact that serum, whether from immunized or nonimmunized animals, contains enough RNase to cause inactivation of viral RNA. However, when the gamma globulin fraction rather than the whole serum is used, unambiguous results are usually obtained: The infectivity in nucleic acid preparations is not lost by treatment with the virus-specific antibody, a strong indication not only that the infectivity is not due to contaminating virus, but also that it is not dependent on the presence of viral protein.

Another useful criterion is the sedimentation rate of the infectivity. Viruses are larger and they sediment much faster than their nucleic acids, and intact nucleic acid molecules show quite characteristic sedimentation rates under standard conditions of ionic strength, etc. Thus infectivity assays on all fractions after sucrose gradient distribution of viral RNA clearly show whether the infectivity is associated with the typical RNA component (about 30 S for many viral RNA's), whether it sediments like a virus (generally in the range 60–200 S) or like a complex of RNA with a little protein. Other methods of differentiation of viruses and their RNA make use of their different solubilities in molar salts (130a, 147).

In the early studies of infectious TMV RNA it appeared important to establish the absence of viral protein beyond the sensitivity of the techniques discussed above. Standard methods of isolation of nucleic acids rarely reduce their protein content below the level of 1%, which in the case of TMV could correspond to a complex of one molecule of TMV RNA with one coat protein subunit. However, if the RNA is prepared in the presence of bentonite, the protein contamination is appreciably less, and in the case of TMV it is further reduced (to 1 protein per 20–50 RNA molecules) by means of the reconstitution reaction (see later). The determination of traces of contaminants is always technically difficult, and the validity of various tests at minimal concentration levels dubious. These particular experiments were performed with ^{35}S as a means of labeling the protein; all that can be inferred from these tests is that the residual "protein," as deduced from ^{35}S counts, would correspond to 0.04% on the basis of the sulfur content of TMV coat protein. Actually, the traces of amino acids detected in such preparations did not correspond in their proportions to those of the coat protein. That preparations containing 0.04% "protein" and others containing 1% did not differ in their infectivity (per mg of RNA) clearly establishes the intrinsic nature of the infectivity of viral RNA (129, 419). For no other viral RNA or DNA has this point been proved so rigorously, but there is no reason to doubt the general validity of the conclusion that the infectivity of viruses resides in their nucleic acids.

For most viruses the infectivity of the nucleic acid moiety can be shown directly by applying the separated nucleic acid to the host cell or tissue, although at times this may require special tricks. In other cases the multiple nature of the nucleic acid complement of a virus is believed to prevent demonstration that the nucleic acid alone is fully competent as an infective agent, but there is no reason to doubt this.

4. *In Vitro Activities of Viral Nucleic Acids*

The term infectivity as used here signifies the total gamut of entry into the host cell, nucleic acid replication, synthesis of viral coat protein(s) and other virus-specific products, maturation of virus particles, and release of virus with or without cell lysis or other cytopathological or generalized disease symptoms. These various biological and biochemical components of the concept of infectivity are discussed in turn in subsequent chapters. However, attention should be drawn here to the fact that two aspects of the biological activity of viral nucleic acids, namely their template and messenger activities, can be studied in systems simpler than the living host cell, and particularly with purified enzymes in soluble *in vitro* systems (322, 318, 171). It then becomes evident that the entire intact nucleic acid molecule or complement or a virus need not be required for such isolated nucleic acid functions as, for example, coat protein synthesis. Thus the use of separate RNA fragments or components can give insight into the functional role of different parts of the nucleic acid molecule or complex. Such studies are referred to in Section XI,B.

I. Complex Viruses

Two well-studied groups of complex viruses are the myxoviruses and the paramyxoviruses. The capsids in both of these groups represent rods or threads of varying lengths, and clearly helical in the paramyxoviruses. They are enclosed in an envelope which gives the virus a somewhat heterogeneous (pleomorphic) but predominantly spherical appearance. The envelope contains not only the cell membrane-related or -derived lipoproteins occurring in all viral envelopes, but also the characteristic neuraminidase and the hemagglutinating and hemolytic proteins. The question of the viral nature of these components has been discussed in Chapter VII. The envelope also carries an orderly array of about 2000 spikes with which the hemagglutinin activity seems to be associated (see Figs. 7-1, 8-4).

Influenza virus strains have been used extensively in experiments on genetic recombination, cross reactivation, and multiplicity reactivation, all tests for the possibility of interchange of parts of the genome between viruses of different strains, whether both are active or one or both are inactivated. In all of these regards, influenza virus has been the only single stranded RNA virus which has given strongly positive results in many laboratories (109). This singular behavior appears to find its explanation in the recent finding that the RNA of the virus consists of several molecules (see Section VI,D). It is now clear how parts of the genome of an inactivated virus particle may, through recombination, become replicated and incorporated in infective progeny virus. The presence in a stable recombinant influenza virus of neuraminidase from one parent, and the other main proteins from the other parent, has been demonstrated (262).

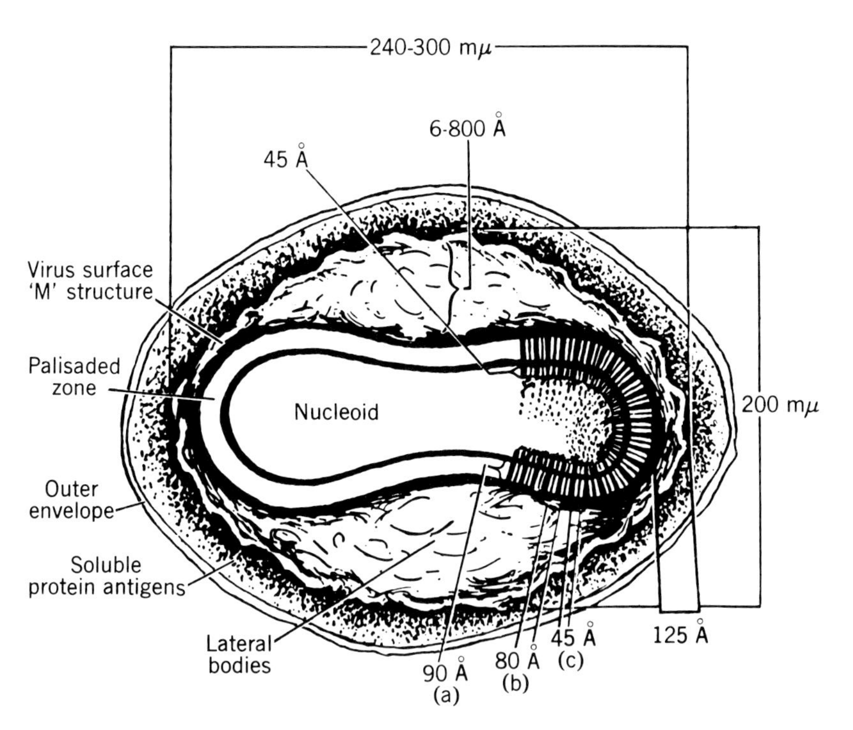

FIG. 8-13. Diagram of the side elevation of a vaccinia virus particle (510). The author wishes to thank Reinhold Book Corporation for permission to reproduce this diagram.

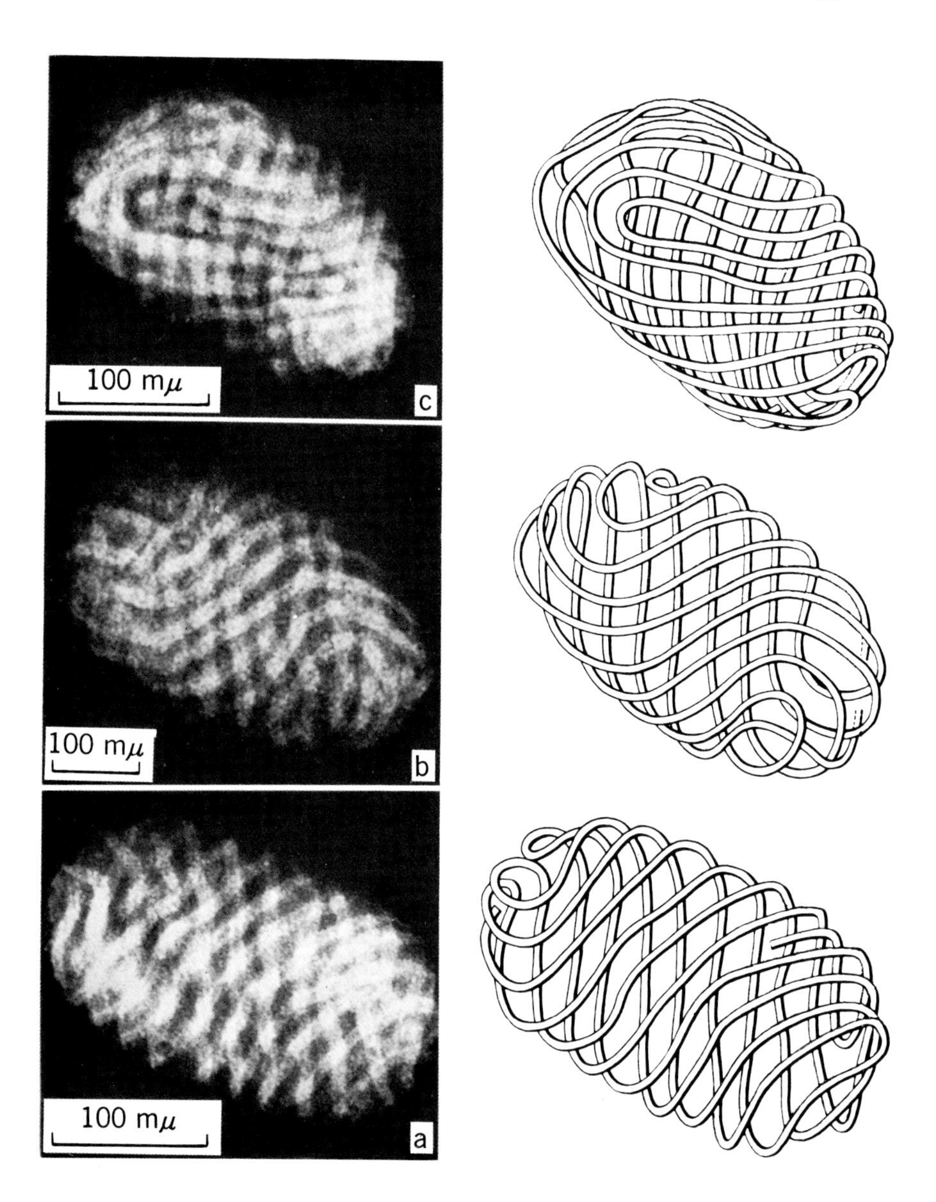

FIG. 8-14. Three particles of Orf virus, negatively stained. Selected to illustrate three ways in which the thread pattern may be seen. The winding patterns have been traced and mounted beside each particle (315).

As stated, the paramyxoviruses share many of the envelope and shape properties of the myxoviruses, but their RNA is much larger than the total complement of myxovirus RNA, and the paramyxovirus RNA definitely represents one molecule (see Table 8.1) (323, 88). A particular biological property of these viruses is their ability to fuse unrelated cells when used at high concentrations. This ability is a useful tool in tumor virus research because it permits infection of normally nonsusceptible cells (254, 255).

The most complex of the viruses in structure are the pox viruses; this is not surprising since they exceed many microorganisms in size. These viruses show such distinctive morphologies that they can be classified on that basis (510, 315). The true pox viruses (vaccinia, myxoma, etc.) are oblong, brick-shaped bodies approximately 100 $\times$ 200 $\times$ 300 mμ in dimensions. The dumbbell-shaped central core presumably contains the nucleocapsid and another structure called the lateral body. Several distinct membranes form layers around these components (see Fig. 8-13).

The paravaccinia group is slightly smaller (160 $\times$ 260 mμ) but appears to have a similar interior structure. In one member of this group (Orf virus) a membrane near the surface looks like a basket weave, owing to the folding of helical protein threads (Fig. 8-14) (315). When this virus is partially degraded, many particles can be seen interconnected by the unraveled threads. Not surprisingly, the intracellular uncoating of this virus represents a special case illustrating mutual adaptation of virus and host.

The structure of the various parts of the tailed bacteriophages might well be discussed here. However, these structures are so closely linked with the process of replication that it is advisable to discuss them in that context (see Section XI,C).

J. Stripping of Helical Viruses

When TMV and, presumably, other helical viruses are treated under carefully controlled conditions with disaggregating solvents (e.g., SDS or urea), the protein is released in soluble form (169). This uncoating can be arrested at various stages before it is complete. It was shown by electron microscopy that such partial

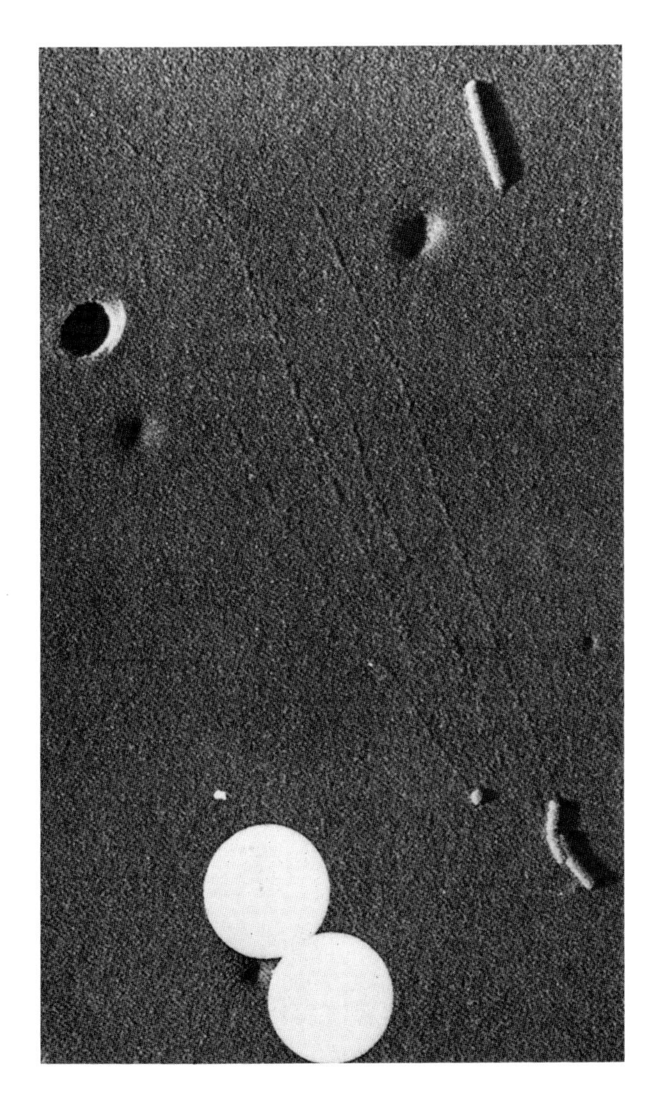

FIG. 8-15. Visualization of the RNA in partially degraded TMV. Note the length of the protruding RNA strand. (From Hart, 169). The polystyrene latex spheres have a diameter of 260 mμ.

stripping did not generally occur in random fashion, but proceeded from an end, yielding shortened rods with a protruding RNA strand (Fig. 8-15). Subsequently it was demonstrated that stripping started preferentially from one end of the rods, namely, the one characterized by the presence of the 3′-glycol (5′-linked) end of the RNA (295). It appears probable that reconstitution and virus maturation also begin at that end.

The technical possibility of denuding a viral RNA in polar and stepwise manner has been put to use for several purposes. One is the location of various cistrons along the RNA chain (see Section VI,A). The treatment of such a partly denuded RNA with a mutagenic reagent which reacts much more readily with free RNA than with encapsulated RNA (e.g., nitrous acids; see Chapter IX) permits the detection of an increased mutability of a particular cistron concomitant with the decapsidation of a particular RNA segment. On such a basis it was concluded that the gene which is responsible for *N. sylvestris* responding with local lesions rather than systemically is located about 1500–2000 nucleotides from the 5′-end of TMV RNA (231). More recent recombination experiments with stripped, and thus shortened, genomes of TMV and complete infective HR have suggested that the coat protein gene is located in the right half (see Table 6.1) of the genome (232).

Stripping has also made it possible to determine the location of certain unique nucleotide sequences lacking G on the RNA chain (see Section VI,A). From such data it was possible to conclude that the coat protein gene is not located near the 3′-end of the RNA.

K. Pseudoviruses

Under this heading are described several unrelated phenomena dealing with noninfectious viruslike particles. The formation in infected cells of viruslike particles lacking nucleic acid is a well-established and widespread phenomenon. A so-called *top component* was first described as accompanying TYMV, and has since been detected in tissue extracts containing many other isometric plant, animal, and bacterial viruses (see Fig. 8-8) (236, 283, 222, 112). In each instance those particles show the same molecular architecture as the complete virus on electron micrographs, but they tend to flatten out on drying and shadowing, and the absence of nucleic acid becomes apparent on negative staining. The top components are separable from typical viruses by sedimentation, particularly on density gradients, but frequently have the same electrophoretic mobility.

The biological role and origin of these nucleic acid-free particles is not clear. Three possibilities have been discussed: Excess protein aggregates to form particles which serve no purpose; protein particles are precusors in virion production; and virus particles can lose their RNA. The second possibility gains some support from the observation that empty phage heads accumulate in infected cells before their combination with DNA and tails (525). Also, the poliomyelitis top component has been shown to play a procapsid role. The maturation of the empty particle to the RNA-filled particle is accompanied in this case by the splitting of a larger protein, characteristic of the top component, into the two typical poliovirus components, VP2 and VP4 (283, 222, 223).

The existence of TYMV particles containing substandard amounts of RNA also suggests that the empty capsids are empty temporarily (293). It was believed until recently that isometric shells could not form spontaneously and, when reconstitution of such viruses was accomplished, it was still surmised that only with and around RNA, or possibily some other core material, could isometric shells be formed *in vitro*. However, it has recently been shown that the protein of the phage fr, dissociated to monomers of 16,000 molecular weight and freed of over 99.9% of the RNA, was able to form viruslike shells on dialysis against 0.1 *M* NaCl at pH 7.8 (180). This evidence favors the first two possibilities discussed above concerning the nature of empty viruslike particles, and the provirus interpretation may now be the preferred one.

Other instances of viruses defective in the sense of containing less than the normal amount of nucleic acid is found in many RNA phage mutants, as well as phages of this group (f2) grown in the presence of fluorouracil, and those reconstituted without maturation factor (see Section X,C). They contain RNA of 14–18 S (instead of 26 S) and are noninfectious while appearing to be identical with the typical phage on electron micrographs. Recent studies with a 14 S RNA mutant of f2 indicates that it lacks the 5′-end and the coat protein cistron of the complete viral RNA (273).

Quite another type of pseudovirus is phage α (alternative names: PBS, χ, μ). This phage is able to lyse the host cell and to attach to a new host cell which is killed as a consequence, but no new phage arises. It has been shown that the failure to produce new phage is due to the absence of viral nucleic acids from these phage particles, for they contain host (*E. coli*) DNA instead. This is true for "most if not all" of these pseudovirus particles (327, 313).

A related phenomenon, but not quite so depressing as the "murder + suicide" committed by phage α, is the presence in polyoma virus preparations of some particles (about 20%) carrying host cell DNA in lieu of phage DNA (465, 520). The amount of this linear DNA (20 S, 3×10^6 molecular weight) corresponds to that of the circular viral DNA (14.6 S) that it replaces. These particles show the hemagglutinating activity, but surely lack the infectivity of polyoma virus. They are pseudoviruses.

CHAPTER IX

THE MODIFICATION OF VIRUSES AND THEIR COMPONENTS, MUTAGENESIS, AND GENETICS OF VIRUSES

A. Reactions

Virus diseases are combatted best by the antibodies formed in most vertebrates in response to viral infection, nature's way of protecting the victim. Since this method represents a cure which frequently sickens and sometimes kills, an important goal of applied virology has always been to find means of decreasing the pathogenicity without affecting the antigenicity of viruses. The use of favorable strains or the *ad hoc* development of attenuated strains, usually by long-repeated passage through suitable hosts or cells, has presented one important path toward that goal. The other path has been to modify viruses chemically so that they become noninfectious but retain their original antigenic properties. This practical purpose represented one of the early stimuli to research on the modification of proteins and, later, also of nucleic acids (432). These fields have been covered in several recent reviews (66, 329, 421). *Methods in Enzymology*, Vol. XI, 1967 (C. H. W. Hirs, ed.), pp. 481-648, also contains a series of reviews of protein reactions, by various authors.

More recently it was realized that the study of the action of certain reagents on proteins, nucleic acids, or complexes of the two can furnish very useful information concerning the primary structure (amino acid and nucleotide sequences), conformation, and function of these macromolecules.

In particular, with respect to the nucleic acids, it is to be expected that modification of those parts of the purines and pyrimidines which are destined to engage in complementary binding would either abolish that binding capability or change it to the alternative form, thus resulting in mutation. Since the binding capability of each base is determined by the great predominance of one tautomeric form over the other (498), reagents which do not directly act on the groups involved may nevertheless have a mutagenic effect by changing the equilibrium between the two forms (421).

The preferred working hypothesis is thus as follows: In the search for reactions which abolish the infectivity of a virus, reagents which prevent base-base interactions, a prerequisite for replication, should be selected. If, on the other hand, the production of mutants is the objective, then reagents which affect the tautomeric equilibrium of one or several bases, preferably by direct action on the involved atoms, is advocated. The study of such mutagenic reactions, and their effects on viruses and viral as well as other nucleic acids, has become a very fruitful and informative line of research.

A logical corollary of the fact that certain modifying reactions prevent base pairing is that base pairing protects against attack by these reagents. Thus, double stranded nucleic acids are remarkably resistant to many reagents which react readily with single stranded polynucleotides (421). Other polymer interactions similarly and selectively protect many groups in proteins and nucleoproteins. In this manner modification reactions can serve also as useful tools in the study of the conformation of proteins and viruses.

The first reagent to be used for the preparation of a viral vaccine (ideally a noninfectious, antigenically unaltered virus) was *formaldehyde* (Table 9.1). It is now known that the primary reaction of formaldehyde is with the amino groups of both proteins and nucleic acids (114, 430, 118, 174, 107, 108). This reaction does not greatly alter the physicochemical properties of proteins (hydrophilic and basic groups remain hydrophylic and basic after addition of the

PROTEINS			NUCLEIC ACIDS
Formaldehyde: $-NH_2 \leftrightarrow -NH-CH_2OH \leftrightarrow N(CH_2OH_2)$			(Same) (A, G, C)
$-NH_2-CH_2-OH + HX \rightarrow -NH-CH_2-X$ $(+ H_2O)$			Same
$HX = CONH_2$, $-NH_2$, tyr, trp, his			
Nitrous acid: $-NH_2 \rightarrow OH$ (lys, N-terminus)			$-N{=}C(-NH_2)- \rightarrow -NH-C(=O)-$
$-N=N-$ (tyr, $-$) (etc.)			(A, G, C)
Hydroxylamine (H_2NOH)			pH 6: C, pH 9: U (see Fig. 9-2)
Acid anhydrides		$-NH_2 \rightarrow -NH-CO-R$	
$R=CH_3$		$-SH \rightarrow -S-CO-R$	
Acetic anhydride	$R=CH_3$	tyr$-OH \rightarrow$ (tyr)$-O-CO-R$	
Succinic anhydride	$CH_2-CO-O-CO-CH_2$ (cyclic)	$-NH_2 \rightarrow -NH-CO-CH_2-CH_2-COOH$	
Maleic anhydride	$CH{=}CH$ with $-CO-O-CO-$ (cyclic)	$-NH_2 \rightarrow -NH-CO-CH=CH-COOH$	
Alkylating agents		$-SH \rightarrow SR$	$G \rightarrow 7RG$
$(CH_3)_2SO_4$, $(C_2H_5)SO_4$ $(R=CH_3, C_2H_5)$		$-NH_2 \rightarrow NHR$	$A \rightarrow 1RA$
CH_3SOOCH_3, $CH_3SOOC_2H_5$ $(R=CH_3, C_2H_5)$		(tyr) $OH \rightarrow OR$	$C \rightarrow 3RC$
ICH_2COOH, ICH_2CONH_2 $(R=CH_2COOH,$ $CH_2CONH_2)$		(met)$SCH_3 \rightarrow {}^{+}S(CH_3)R$	$A \rightarrow 7RA$
$(R=CH_2CH_2OH$		(his)$NH \rightarrow NR$	$A \rightarrow 3RA$
CH_2-CH_2 (epoxide, O) $CH_3-CH_2-CH_2$ (epoxide, O)		$CH_2CH_2OHCH_3)$	(see Fig. 9-3)
$Cl-CH_2CH_2S^{+}(CH_2)_2$ (cyclic) Cl^-		$(R=CH_2-CH_2-S-CH_2-CH_2Cl)$ cross linking	
"Nitrosoguanidine" $(H_3C)(ON)N-C(=NH)-N(NO_2)H$			Same

reagent), nor does it markedly affect their antigenic properties. On the other hand, the same addition of -CHOH groups to the amino groups of the purines and pyrimidines destroys their ability to function as templates or messengers, and this amounts to inactivation of the virus. In turn, formaldehyde does not react appreciably with DNA, unless its hydrogen bonds have first been broken by denaturing agents. Formaldehyde treatment is not expected, nor found, to be very mutagenic. It would seem that the first reagent to be studied fully attained the objective of preparing a perfect vaccine from RNA-containing viruses. The Salk polio vaccine was based on this reaction.

However, a disadvantage of aldehyde-amine interactions is their reversible nature. Thus infectivity may reappear when the excess aldehyde is removed or diluted. Furthermore, the aminomethylol groups are quite reactive and tend slowly to form more or less stable methylene crosslinks, in both proteins and nucleic acids, through secondary condensation reactions with various other groups (118, 109) (Table 9.1). Thus some of the reagent gradually becomes firmly and irreversibly fixed to the polymer. These effects tend to alter the antigenic nature of a protein, as well as the physicochemical nature of both biopolymers.

Another reagent which has long been in use in protein chemistry is *nitrous acid.* Its primary reaction is oxidative deamination, i.e., replacing amino by hydroxyl groups both in proteins and in nucleic acids (393, 395). Again, double stranded nucleic acids are quite unreactive in principle, but the pH of 4.5 or less required to maintain a sufficient HNO_2 concentration (NO_2^- is not active) causes incipient denaturation, particularly releasing the guanine-amino groups, and their deamination further weakens the double strandedness of the molecule (421).

As illustrated in Figs. 6-4 and 9-1, the deamination of cytosine and adenine to uracil and hypoxanthine, respectively, results in tautomeric shifts and thus in direct mutagenesis. The transformation of guanine to xanthine is inactivating rather than ineffectual because the new —C = O group seems to repel the —C = O group on cytosine. Nitrous acid has been the most useful reagent in the elucidation of the genetic code and for the production of mutants, and it is still the only reagent which permits transformation of one

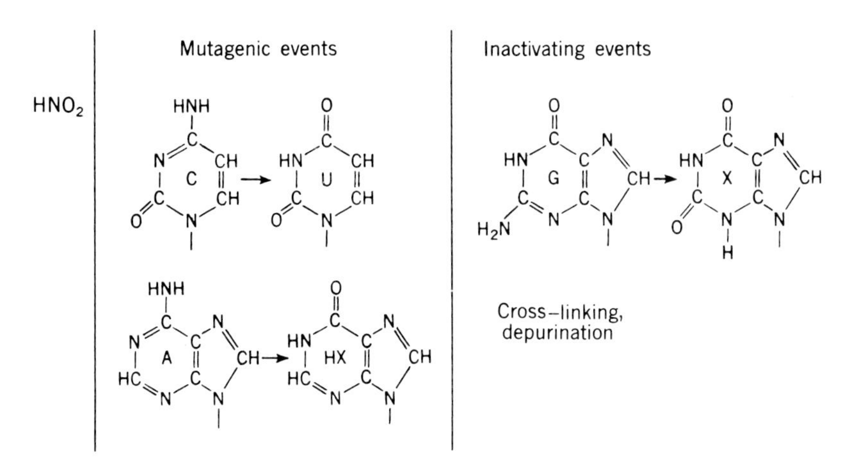

FIG. 9-1. The reaction of bases with nitrous acid. The deamination of guanine is not necessarily always inactivating. The cause for mutagenesis in terms of hydrogen bond donating and accepting capabilities is also illustrated in Fig. 6-4.

typical RNA component into another, cytosine into uracil. As for formaldehyde, a disadvantage is lack of specificity (three bases react), and the occurrence of side reactions. Proteins quickly become yellow and pink through reactions involving the tyrosine and tryptophan residues and leading to crosslinking. In nucleic acid, particularly in DNA, crosslinking and depurination resulting from the acid pH complicates the interpretation. Also, the reaction with the guanine residues is complex, and xanthine not the only product (403).

Hydroxylamine does not react readily with any typical protein group, but it does react readily with the pyrimidines of nucleic acid. The uracil ring is opened with a pH optimum of 9. Near neutrality, and particularly at pH 6, two reactions with cytosine predominate greatly (394, 265). They are: the direct replacement of the amino by the hydroxyamino group (Fig. 9-2, reaction a); and the addition of the reagent to the 5,6-double bond, followed quickly by the replacement of the amino group by a hydroxylamino group (Fig. 9-2, reactions b, c). The action of hydroxylamine on TMV RNA is highly mutagenic, and this is attributed to reaction a (421), since the preferred tautomeric state of that compound is that of an

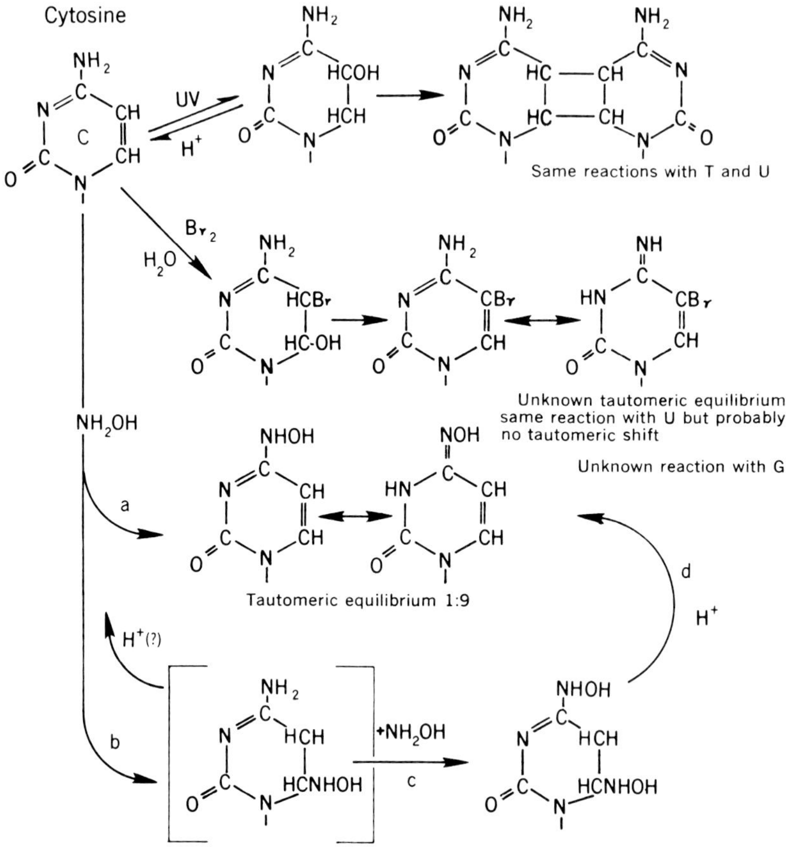

FIG. 9-2. The reactions of UV light, bromine, and hydroxylamine with pyrimidines. The mutagenic action of bromine and hydroxylamine is attributed to tautomeric shifts to the products on the right in the two middle rows (49). The ring-opening action of hydroxylamine on uridine at pH 9 is not shown.

oximino uracil (49, 50). Whether the products with saturated double bond are inactivated is not yet established.

Attention might be drawn here to the fact that all reactions which are mutagenic are of necessity also inactivating, since many of the mutational events must represent nonsense mutations (terminating the peptide chain during translation; see Section XI,B), or yield, through amino acid replacement, nonfunctioning proteins.

The disadvantages of hydroxylamine are: double stranded DNA does not become mutated or inactivated unless denaturing conditions are used; the RNA in certain virus particles, particularly TMV, is also quite unreactive, for unknown causes; hydroxylamine tends to decompose, yielding radicals which cause inactivation by unspecific reactions that are difficult to control (421).

In terms of searching for the best agent for the production of vaccines, the high mutation rate of nitrous acid and hydroxylamine treatment is regarded as a serious disadvantage because of the fear of producing a superactive mutant. In actuality, of the many hundreds of artificial mutants of TMV produced since 1958 (see Section IX,B), none has been found to be of unusual virulence, and almost all are less infective, productive, and/or stable than the wild type. This presumably is a consequence and illustration of the fact that evolution has been at work selecting the fittest viruses for a sufficiently long time to make marked improvement, or increase in virulence, improbable.

Acylating agents (acetic, succinic, or maleic anhydride, benzoyl chloride, etc.) have long been used in protein chemistry, but they do not modify nucleic acids in aqueous solutions near neutrality. Various acylation reactions have been used recently as probes into the conformation of the TMV protein in the virus. Succinylation and the readily reversible maleylation reactions have proved to be useful tools for the solubilization of virus proteins by changing the —NH_3^+ groups to —COO^- groups and thus causing the subunits to dissociate (52, 139, 411). The amino, phenolic, and thiol groups can become acylated. Of these groups, generally only the amino groups are on the surface of proteins and thus these reactions appear rather amino-specific. In particular cases, however, phenolic groups can react as readily as amino groups, and unprotected —SH groups are very readily acylated. In TMV, one of the four tyrosines (the one nearest the C-terminus, #139) is more reactive than one of the two amino groups (#53) (see Fig. 5-7a) (123, 335).

Alkylating agents (dimethylsulfate, diethylsulfate, iodoacetate, ethylene and propylene oxide, sulfur or nitrogen mustard gas type compounds, ethyl and methyl methane sulfonate, etc.) react with proteins and nucleic acids, but probably more readily with the latter. In proteins, available thiol, amino, and phenolic groups become alkylated, more or less in this order, although through variation in pH the reaction can, in a given protein, be directed toward one or the other of these groups (in the absence of by far the most reactive unmasked —SH groups), or toward methionine residues, or toward a particularly and unusually reactive histidine or carboxyl group. A class of reagents can hardly be less specific and yet, through careful selection of reaction conditions, more useful (Table 9.1).

In nucleic acids the 7 position of the guanine is by far the most readily alkylatable site, followed by the 1 position on adenine and the corresponding 3 position on cytosine (264, 120). Other alkylations of A and alkylations of U occur under certain conditions (see Fig. 9-3).

The preference of alkyl radicals for a site on RNA and DNA which is unprotected by base pairing (the 7 position of the guanines) differentiates them from all previously discussed reagents. Also, the packaging of the RNA in the TMV rod, which affects the reactivities of the amino groups of adenine and particularly of guanine, does not interfere with the alkylation of the 7 position of guanine. Thus a powerful tool would seem to be at hand but for the ambiguity of the effect of such alkylations. The preferred

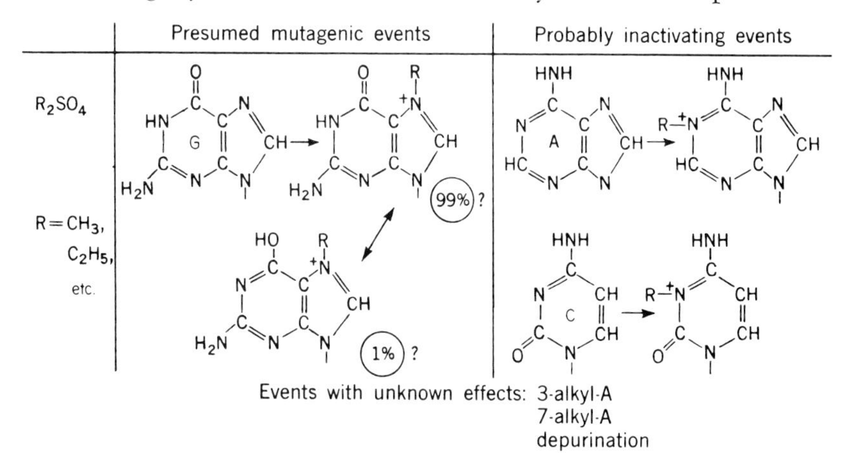

FIG. 9-3. The reaction of bases with alkylating agents. The tautomeric equilibrium of 7-alkylguanine derivatives is unknown, and the 1% figure only a guess. The alkylation of adenine and cytosine in the 1 and 3 positions, respectively, as shown, are believed not to be mutagenic because they would be expected to interfere with base pairing.

working hypothesis (421) for the biological effects of alkylation is as follows: The substitution of the 7 position of guanine by small alkyl groups ($-CH_3$, $-C_2H_5$) is not inactivating, but introduction of large substituents, particularly two-pronged ones that produce crosslinks (such as mustard gas), causes inactivation. The effect of 7-alkylation slightly favors the unusual tautomeric form of guanine, so that one out of possibly a hundred 7-methylguanine residues may code like adenine, and thus such alkylations may occasionally be mutagenic (Fig. 9-3).

The main site for alkylations of adenine (N 1) and the only alkylatable site of the pyrimidines (N 3) involves the base-pairing part of the molecules, and therefore these reactions probably occur only in single stranded nucleic acids and are inactivating. The summation of these reactions means that, under nondenaturing conditions, treatment of double stranded nucleic acids with alkylating agents produces many low level mutagenic events not competed for by any inactivating events. Mutagenic rates must always be evaluated in relation to inactivation rates, and thus alkylation of DNA is highly mutagenic. In RNA, in contrast, inactivating reactions do occur, and alkylation is therefore poorly mutagenic.

One additional fact remains to be explained: TMV RNA is mutated slightly by dimethylsulfate but hardly at all by diethylsulfate and most other alkylating reactions, whereas DNA is mutated most effectively by ethylation. The interpretation touches on one of the side reactions of alkylation. The substitution of the N 7 of guanine renders this a quaternary (strongly basic) nitrogen, and thus weakens the glucoside linkage on N 9. The deoxysugar-purine bond is notoriously weak (see Section VI,A) and, if further weakened by N 7 alkylation, it tends to break even at neutrality. It seems that methyl substitution weakens that bond more than does ethyl substitution. Thus 7-methyldeoxyguanine may be released as the base almost as fast as it is formed, and methylation of DNA may therefore not be highly mutagenic.

N-Methyl-N′-nitro-N-nitrosoguanidine (MNNG, *nitrosoguanidine*) is a very effective bacterial mutagen, the chemical effects of which on viral nucleic acids were not studied intensively until very recently (420, 63, 422). The main action is alkylation, which yields the same products as typical alkylating agents. However, MNNG differs in a number of respects from typical alkylating agents, as follows: (a) MNNG alkylates guanine much more readily in polymers than in the form of free guanylic acid or other small molecular compounds. (b) MNNG reacts very poorly with the guanine in the TMV rod [in contrast to dimethylsulfate (DMS), which reacts readily], and in dispersing solvents such as formamide. The conclusion from these data is that the reaction of guanine with MNNG requires the base-stacked array which is the characteristic state of nucleic acids in aqueous solution. The reagent probably intercalates between bases before it can transfer its methyl group. (c) MNNG is a poor mutagen similar to DMS, when acting on TMV RNA (and DNA?) under conditions where the guanine reaction is favored. It is a better mutagen in formamide and a very good one in the virus rod. It appears unlikely that the high mutagenic action of MNNG on intact TMV can be attributed to its ability to alkylate guanine. Whether this mutagenicity is due to any methylation reaction or, possibly, to a deaminating reaction (422) remains to be established. The high mutagenic activity on *in vivo* treatment of phages and bacteria with MNNG is observed at much lower concentrations than required for the reactions with isolated viruses or nucleic acids, and it seems probable that under these conditions mutagenesis results from an action of the reagent at the growing point during DNA replication (see later) (62).

Ultraviolet light is frequently used for the inactivation of viruses (303). Proteins containing only small amounts of U.V.-absorbing components are very much less sensitive to this agent than are nucleic acids. Both single and double stranded nucleic acids become inactivated, but the mutagenic rate is generally low. The first chemical events resulting from intense UV irradiation are the addition of water to the 5,6-double bond of the pyrimidines—a reaction which is reversible with different ease for different pyrimidines (similar to the addition of hydroxylamine to this bond) (399). Subsequently dimers form between neighboring pyrimidines (34), a much more stable and surely inactivating reaction (see Fig. 9-2). The reaction of viruses with UV light is more complex than that of the nucleic acids alone. Different strains of TMV show great differences in their sensitivity, and evidence has been obtained that protein-nucleic acid interactions occur during irradiation, although this does not seem to play a significant role in the inactivation kinetics (442).

UV irradiation would appear to be a promising agent for vaccine formation, were it not for the existence of photoreactivation and related phenomena. Quite apart from the ease of chemical reversal of the first step, water addition, there also exist biological mechanisms in most cells, some light-activated, for the repair of nucleic acids damaged by UV and other agents. It has been shown that this repair can consist of enzymatic splitting of interdimer bonds. In other systems the pyrimidine dimers are removed by excision of the damaged area on one strand and rebuilding of the missing segment (the other strand serving as template) by the appropriate polymerases and sealases (see Section IX,B) (399, 328).

Halogens (e.g., Br_2) may also be used to modify and mutate viral nucleic acids. Cytosine is the most reactive base, although uracil and guanine also react with halogenating agents. The first reaction product of the pyrimidines is the result of addition of BrOH to the 5,6-double bond, followed by dehydration to the 5-bromo derivative, which can be further brominated (43). The formation of 5-bromocytosine may result in mutation by favoring the tautomeric shift (see Fig. 9-2).

B. Chemically Produced Mutants

As noted in the preceding section, two modifying reactions are known which clearly change the bonding capability of individual nucleotides in single stranded polymers. Thus messenger nucleic acid molecules interacting with hydroxylamine (at pH 5–6) or nitrous acid (at pH 4–5) to the extent of only a few events are likely to remain unchanged, except for 1–3 cytosines and/or (in the case of nitrous acid) adenines showing the binding properties of uracil and guanine, respectively.*

* The two types of mutational events frequently discussed are transitions and transversions. Transitions correspond to the only type of direct mutational events readily comprehensible in molecular terms and discussed above (6-aminopurine ↔ 6-oxopurine, 3-aminopyrimidine ↔ 3-oxopyrimidine, in other words, something coding like (s.c.l.) A ↔ s.c.l. G, and s.c.l. C ↔ s.c.l. U). Transversions postulate the replacement of a purine by a pyrimidine, and vice-versa, events for which there exists no definite chemical evidence.

These reactions produce the simplest and most clear-cut types of mutational events. When such a nucleic acid becomes translated into amino acid sequences, it is possible, on the basis of the now well-established codon dictionary (Fig. 9-4; Table 5.1) to predict which

First base	Second base U	C	A	G	Third base
U	PHE	SER	TYR	CYS	U
	PHE	SER	TYR	CYS	C
	LEU	SER	TERM	TERM	A
	LEU	SER	TERM	TRP	G
C	LEU	PRO	HIS	ARG	U
	LEU	PRO	HIS	ARG	C
	LEU	PRO	GLN	ARG	A
	LEU	PRO	GLN	ARG	G
A	ILE	THR	ASN	SER	U
	ILE	THR	ASN	SER	C
	ILE	THR	LYS	ARG	A
	MET,F-MET	THR	LYS	ARG	G
G	VAL	ALA	ASP	GLY	U
	VAL	ALA	ASP	GLY	C
	VAL	ALA	GLU	GLY	A
	VAL	ALA	GLU	GLY	G

FIG. 9-4. The codon dictionary.

amino acids may occasionally be replaced by which.

The effect of nitrous acid on *TMV and TMV RNA* has been under study since 1958 (393, 148), i.e., before elucidation of the code. It is clearly evident that this reaction produces many mutants in terms of changing the character of the systemic symptoms evoked by individual TMV particles and their progeny on *N. tabacum*, or the systemic or chlorotic to a necrotic lesion response on other varieties of tobacco.

When the progeny of such mutated virus particles is isolated and its coat protein is analyzed, one, two, or, rarely, three amino acid replacements are observed in 30–50% of the cases (473, 523, 143). That the coat proteins of many mutants remain unchanged in amino acid composition,* and probably also in sequence, is not surprising, since only 8% of the message of TMV RNA codes for the coat protein (see Section XI,B). Amino acid replacements in other gene products (e.g., the RNA replicase) cannot yet be analyzed for in this system. In fact, replacements in the coat protein occur much more frequently than might be expected, possibly because biologically detectable mutations may require multiple hits in one or several cistrons, and coat protein alterations may be coincidental, though surely often lethal, by-products.

When the nature of the amino acid exchanges is considered, it becomes apparent that most of them correspond to the expected results for deamination of cytosine and adenine. Prolines (CC^{Pu}_{Py}) are often lost, but never phenylalanine (UU^{Pu}_{Py}), and all the most frequent replacements (Pro → Leu or Ser; Ser → Phe or Leu; Thr → Ile, Ala or Met; Ile → Val; Asn → Ser; Arg → Gly) can be accounted for by a cytosine-to-uracil or adenine-to-hypoxanthine change in the codon. There are exceptions, but they occur singly and may well be due to spontaneous mutational events, unrelated to the action of nitrous acid. On the other hand, a number of expected exchanges (e.g., Ala → Val; Lys → Glu or Arg) have never been observed, and several of them cannot readily be explained on protein-chemical grounds. In a very few instances two nucleotide replacements must be assumed to account for a given replacement (e.g., Ser → His, AGPy → UC^{Pu}_{Py}), also a highly unexpected event (360). (Pu and Py stands for either purine or pyrimidine.)

In mutants showing more than one amino acid replacement, the replacements are never adjacent, quite in contrast to a group of T4 mutants to be discussed below.

The only absolute proof for the identity of two proteins is the

* Much more rarely is a major change from class A to B or C detected (see Table 5.2); the significance of such an event is not yet clear (122). As far as the mutants showing wild type composition are concerned, it should be recalled that a natural strain of TMV (the masked strain, giving no symptoms on *N. tabacum*) also has the same amino acid composition, and presumable sequence, as the wild type.

determination of their complete amino acid sequence, since amino acid analysis alone would fail to detect pairs of compensatory exchanges (e.g., Phe → Leu in one part of the molecule and Leu → Phe in another). The earlier statement that many TMV mutants showed unchanged coat proteins is therefore subject to question, if it is based only on amino acid analyses and not on complete sequencing. As implied by the preceding discussion, however, nitrous acid mutagenesis yields theoretically (and almost always in practice) only monodirectional exchanges (Ser → Phe, not Phe → Ser, etc.). Thus compensatory replacements are not permitted, and all exchanges would have to lead to analytically detectable differences. Only in one instance, when an "improper" exchange had occurred, was a net single exchange actually the result of two exchanges one of which was compensatory (Gly → His = Gly → Ser (GGPy → AGPy) + Ser → His, AGPy → UC^{Pu}) (360).

The consequence of mutational events in terms of protein chemistry and function is obviously often deleterious or destructive. As stated elsewhere, all TMV coat protein mutants that have been isolated are less stable than the wild type. Many more were too unstable for isolation of the virus. On infection by a few of the defective strains, proteins were isolated from leaf tissue, and they retained the ability to aggregate in helical manner but were unable to form rods with TMV RNA (wild type or mutant). When two of these proteins were analyzed (413, 414, 540), they showed, respectively, one and two exchanges, two of which were at the edge of "immutable" regions (residues 95 and 112; see Fig. 5-7).

The only other viruses which have lent themselves to similar evaluations of mutagenic events are the *RNA phages of the f2 class*, since these, like the TMV group, have coat proteins of known amino acid sequence. On growth in the permissive host S26R1E (su I), several defective mutants of R17 obtained with nitrous acid were found to have a serine in place of one of three glutamine residues (residues 6, 50, 54). The presumed mutational event is thus CAG (glutamine) to UAG (amber), the latter being read as if it were serine (UCG) by the sRNA of the permissive su I host strain (see below) (468, 408).

Attention has been focused on nitrous acid mutagenesis, although hydroxylamine was mentioned as possibly the most specific direct mutagen. Yet, for no evident reason, hydroxylamine mutants of TMV RNA have shown almost no amino acid replacements in the coat protein, nor does this reagent seem to have been used as a mutagen with RNA phages.

The mutagenic procedures most frequently employed actually do not lend themselves to chemical evaluation, since they allow the mutagen to act on living cells and replicating nucleic acids. Thus mutagenesis is most frequently due to complex biochemical events rather than to a definite modification of a codon. The same type of mutants of R17 RNA as produced by nitrous acid (*in vitro*) and attributed to cytosine-to-uracil exchanges are also obtained by growing the virus in the presence of 5-fluorouracil (0.0001 *M*) (468, 408, 200). Fluorouracil is very similar in its physicochemical properties to uracil (since the atomic radius of F does not differ greatly from that of H), and it is readily incorporated during biosynthesis of any RNA. TMV RNA with half of its uracil replaced by fluorouracil shows full infectivity and negligibly few mutants (151, 421). No mechanism can be envisaged by which the introduction of fluorine in the 5 position would alter the base pairing of uracil residues, nor have errors of translation been detected with fluorouracil-containing polynucleotides (157). On the other hand, if fluorouracil became metabolically converted into fluorocytosine (this appears quite possible) and as such replaced a few cytosines in a messenger, this might well cause a tautomeric shift and make what is normally cytosine code like uracil and result in glutamine → amber (CAG → UAG) mutations.

As stated, mutagenic agents are of two principal types: chemically active agents which modify nucleic acids *in vitro*, and base analogs which act as mutagens only on being incorporated *in vivo*. Even some chemically active reagents (e.g., nitrosoguanidine) are frequently used *in vivo*, because they are more effective in that manner, and the required reaction conditions are not grossly detrimental to life. Although *in vivo* mutagenesis is very successful in terms of mutant production, such procedures have not greatly advanced the understanding of mutagenesis. It would seem that such understanding can come only from the study of modifications of single stranded nucleic acids, or polynucleotides, and of the effect of well-understood reaction on the messenger or template activity of such nucleic acids.

The restriction of such a study to single stranded nucleic acids is advisable for two reasons: double stranded nucleic acid is quite resistant to the best known mutagenic reactions

$$-NH_2 \begin{cases} \rightarrow =O \\ \rightarrow =NOH \end{cases}$$

so that the results obtained are frequently due to side reactions (421); and double strandedness complicates the interpretation of the data, because the observed effect may be either the direct one or the complementary one. Thus the unambiguous aspect of the reaction is lost.

Besides the extensively used mutagenic base analogs such as fluorouracil, bromouracil (272) (replacing thymine, the atomic radius of Br resembling that of CH_3) and 2-aminopurine, there exists another group of mutagens effective only *in vivo,* the acridine dyes. These dyes have produced many mutants of T-even phages which have lent themselves to interesting molecular analysis.

Proflavin and other acridines are known to intercalate between the bases of double stranded DNA, and there is good evidence that the mutagenic action of such agents consists of deletion and insertion of one or several nucleotides. The loss of over 20% of the length of λ in a double deletion mutant was detected electron microscopically (279). Mutants of this type have provided important genetic evidence for the triplet nature of the code (71, 70). In other experiments double mutations in the same cistron were particularly useful. If a second mutational event restores the loss of a functional gene product (such as lysozyme activity), a deletion of a nucleotide in the RNA may be assumed to be compensated by a nearby insertion (461).

Phage lysozyme has been selected as a gene product of comparative simplicity for protein-chemical investigations of this type. The complete amino acid sequence of T4 lysozyme is now known but, even before this task had been accomplished, it was possible to detect and analyze altered peptide sequences in mutant lysozymes after tryptic degradation. It has become evident from these studies that a group of consecutive amino acids is changed in such a mutant in line with the expected mechanism of proflavin mutagenesis. Such "frame-shift" mutations have supplied impressive illustrations of the correctness of the code dictionary and of the direction of translation, as well as of the ability to predict molecular events on the basis of genetic data. When the first altered lysozyme peptide sequence to be elucidated, and an equivalent back-translated nucleotide sequence (suitably selected from the codon dictionary) are compared with the corresponding ones of the wild type, it becomes evident that the deletion of an adenine and insertion of a guanine 15 nucleotides (5 amino acids) away accounts fully for the observed changes in the protein structure (Fig. 9-5) (461). Among

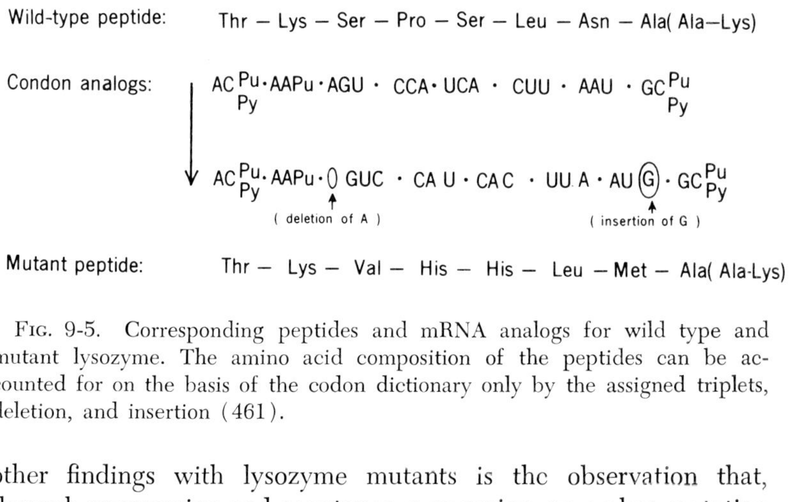

Wild-type peptide: Thr — Lys — Ser — Pro — Ser — Leu — Asn — Ala(Ala—Lys)

Condon analogs: AC$^{Pu}_{Py}$·AAPu·AGU · CCA·UCA · CUU · AAU · GC$^{Pu}_{Py}$

↓ AC$^{Pu}_{Py}$·AAPu·() GUC · CA U · CA C · UU A · AU (G) · GC$^{Pu}_{Py}$

(deletion of A) (insertion of G)

Mutant peptide: Thr — Lys — Val — His — His — Leu — Met — Ala(Ala-Lys)

FIG. 9-5. Corresponding peptides and mRNA analogs for wild type and mutant lysozyme. The amino acid composition of the peptides can be accounted for on the basis of the codon dictionary only by the assigned triplets, deletion, and insertion (461).

other findings with lysozyme mutants is the observation that, through suppression and spontaneous reversion, an amber mutation resulted in a replacement of tryptophan residue 158 by tyrosine (UGG → UAG → UAPy) (217).

C. VIRAL GENETICS

The preceding discussion made frequent reference to viral genetics. The field of viral genetics is now one of the most actively pursued areas of biology. Because it is the topic of many recent books and reviews, no attempt is made here to duplicate this effort

(177). Only a few of the main principles are mentioned in the hope of facilitating the reader's understanding of the molecular aspects of mutagenesis and genetics discussed.

The key to the field of viral genetics was the study of the result of multiple bacteriophage infection. It appeared that related phages could replicate simultaneously in the same cell. The important advance came from the use of related phages differing by two (or more) genetic traits, for these infections led to four (or more) types of progeny phage, a result that could be accounted for only by genetic recombination (80, 181, 182). The principle of classical genetics, the interchange of parental genetic materials, thought to require sexual modes of reproduction, was thus extended to the phages, as it has since been extended to bacteria.

It is now clear that crossing over of viral genetic material, the molecular basis of these phenomena, can and does occur as long as there is free progeny DNA from two strains in an infected cell. The frequency of appearance of specific recombinants represents a measure of the physical distances of the respective genes on the DNA molecule, and it is the basis for construction of genetical maps. As expected, the farther two genetic markers are apart, the more frequently they are segregated. The exceptions in the case of the T-even phages presented the first evidence for the circularity of the genetical, as contrasted to the physical, map of those viruses (see Section VI,D) (98).

The same principles, although not the same operational advantages, hold for RNA phages, and for animal viruses, whether they are RNA- or DNA-containing, to the extent that multiple infection and progeny interchange are possible in the host cell (68). The reasons why recombination frequencies are very much higher with the myxoviruses than with any other viruses have been discussed previously. Only plant virus systems have resisted genetic analysis, probably because of the near impossibility of obtaining multiple infection in plant cell populations.

An important experimental tool for the study of genetical maps which has been touched on in passing is the use of conditional lethal mutants. Thus mutants which are viable at a lower but not a higher temperature have proved extremely useful. In some instances such temperature sensitivity could be traced to the lesser stability of a particular gene product protein, as for example in the case of certain nitrous acid mutants of TMV, which have coat proteins that denature at unusually low temperatures (225).

Besides the temperature-sensitive mutants (ts), there exist many suppressor-sensitive (sus) mutants, also called host-sensitive, among the conditional lethals. These mutants are able to grow on some but not on other host strains. It is now clear that in some instances such mutants (named amber mutants, for no sufficient reason) are due to the presence of unusual tRNA species in the permissive bacterial strain. Thus a mutation may result in the codon UAG, which calls for termination of the peptide chain (a codon called, unfortunately, a nonsense codon). However, owing to the presence of unusual species of tRNA, in the permissive bacterial strains su_{I}, su_{II}, and su_{III}, this codon may be read as if it represented serine (UCG), glutamine (CAG), or tyrosine (UAPy). It is clear that amber mutants are not confined to a particular gene, since they are defined only in terms of nucleotide replacements, leading to UAG triplets. Thus the number of potentially viable mutants available for study is greatly increased, and the genetic mapping of viral nucleic acids by crossing-over techniques is correspondingly furthered (535, 178). Another group of bacterial strains which favor occasional misreading of the codon UAA has allowed the detection of the so-called ochre mutants (444).

CHAPTER X

THE ASSEMBLY AND RECONSTITUTION OF VIRUSES

A. General Remarks

Viruses can be dissociated into their covalently linked components by a variety of agents which weaken or break the ionic and hydrophobic interactions and the hydrogen bonds that hold them together. Infection of animal, plant, and bacterial cells by many of the simpler viruses leads to the formation not only of complete viruses, but also of nucleic acid-free, viruslike particles. Thus the shape of these virus particles, whether rodlike or isometric, is not dependent on the presence of the nucleic acid, but represents a specific protein aggregation phenomenon (see Section VIII,B). It might thus be predicted that, under favorable conditions, viral proteins would aggregate to viruslike particles even *in vitro*. This was first observed with TMV protein obtained from infected cells (457) or from the virus (390).

Complete virus particles containing the nucleic acid are generally much more stable than their nucleic acid-free counterparts, and there would thus be a tendency to predict that *in vitro* formation of complete virus particles would also be possible when viral protein was mixed with viral nucleic acid and the two allowed to interact under favorable conditions. This also is a *post facto* prediction, since the first virus reconstitution, that of TMV, was reported in 1955 (128), and the self-assembly of a number of quite different viruses has been studied more recently.

We now discuss the requirements and conditions for virus reconstitution, the extent to which the viruses produced *in vitro* and *in vivo* are identical, and some of the applications of the reconstitution reaction in studies of the nature of virus infection. We consider first the only well-known reconstitution system, that of TMV, but it can be assumed that many of the conclusions arrived at for this system have general validity.

B. Reconstitution Reactions with TMV Components

The isolation of virus proteins generally requires the breaking of many intersubunit and protein RNA bonds and, since they are of the same type that gives the subunits their conformation, all virus proteins are more or less denatured in the process of isolation. It is now well known, however, that the native conformation of proteins is a function of their amino acid sequence, and that denatured proteins tend to reform their native conformation spontaneously under proper conditions (7). Similarly, newly biosynthesized peptide chains (see Section IX,B) fold automatically into their proper native protein conformation.

TMV protein prepared by the 67% acetic acid method of splitting the virus is in a semidenatured state, and that prepared with 8 *M* urea or particularly with 5 *M* guanidine hydrochloride at neutrality is thoroughly denatured, but native TMV protein can be recovered from each of these media by careful, and sometimes stepwise, removal of the denaturant. One important precaution is the protection of the unmasked —SH group against autoxidation, particularly in neutral or alkaline solution. If the protein solution appears water-clear near pH 7.5, the protein is largely native and dissociated into small complexes, predominantly trimers (see below). That in this state the protein is able to aggregate to viruslike rods can be demonstrated by adjusting the solution gradually to pH 5 when it becomes opalescent and, in the electron microscope, shows rods of TMV diameter, but of many lengths, including much longer ones than are characteristic for TMV (300 mμ).

If to enough native TMV protein to give a final concentration of 1 mg/ml are added TMV RNA (to 0.05 mg/ml final concentration) and a buffer of about pH 7–7.5 (to 0.1 *M* final concentration; pyrophosphate is most advantageous), and the solution is allowed to stand for several hours at 25–37°, rods are formed which are predominantly 300 mμ long. Such solutions show about half as much infectivity as would be expected if all the added RNA had been coated and thus transformed into virus (125–127). This represents a 50% yield of virus reconstitution. Higher yields are rarely obtained, partly because TMV RNA usually contains 20–40% of fragmented molecules. The yields, always calculated in terms of the RNA added, are somewhat increased by using an excess of protein (e.g., 1 mg/ml of protein to 0.01 mg/ml of RNA). When the same mixtures are held at 0°, no rods are formed and the infectivity remains at, or falls below, the low level ($\sim$ 0.05%) shown by the added nucleic acid. The renaturation of TMV protein and its aggregation to the typical helical rod is illustrated schematically in Fig. 10-1 (125, 126).

Several aspects of these reactions require further comment. The protein, as stated above, aggregates at or below pH 5 to rods of any length; however, if the pH of the solutions is brought back to neutrality, the rods are dissociated and a clear solution is again obtained. This reversible aggregation can also be observed by raising and lowering the temperature of TMV protein in 0.1 ionic strength phosphate buffer of pH 6.5 from 5 to 25°; this is an indication that aggregation is an endothermic reaction (260).

In contrast, the reconstituted virus rods are stable over the same pH range as natural TMV (pH 2–10), and they are unaffected by temperature changes up to 65 or 70°. Thus the coaggregation of RNA and protein gives the structure a much greater stability than that of the protein particle. The length of the rod that can be stabilized by the RNA is naturally a function of the length of unbroken TMV RNA (2×10^6 molecular weight), and it can be calculated

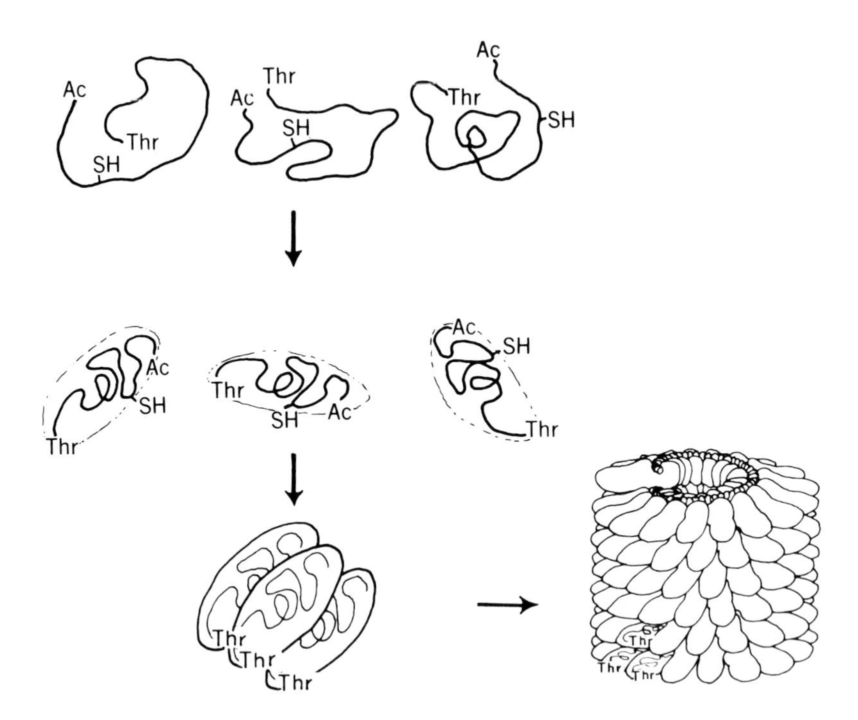

FIG. 10-1. The renaturation of TMV protein and its reconstitution to rods. The spontaneous reformation of a protein of definite shape from its random coil (denatured) state, and the aggregation of such protein molecules to a trimer and to a rod is illustrated.

that 300 mμ is the length of such an RNA molecule forced into the helical groove of 8 mμ diameter and 2.3 mμ pitch which is formed by the aggregated protein subunits.

Other nucleic acids can replace TMV RNA with more or less efficiency in this reaction, including synthetic polynucleotides, as long as they contain a high proportion of purines (127). Of particular interest is the formation of rods predominantly of 150 mμ length when MS2 phage RNA, which is about half as long, is incorporated into TMV protein instead of TMV RNA (448). The proteins of TMV strains can be substituted for that of common

TMV, but the greater the number of differences in amino acid composition and the more distant the relationship (see Section V,B), the poorer is the yield and the more unstable the resultant virus (125, 198). It has been mentioned previously that HR, one of the most distant known relatives of TMV, is much less infective than TMV, whereas the RNA's of the two strains are similarly infective. Now this statement can be added: HR RNA reconstituted with TMV protein is as infectious as reconstituted TMV RNA.

Similar results are obtained when the ability of different strain proteins to coaggregate to protein rods at pH 5 is compared. It appears that the proteins of members of classes A to C can form mixed rods, but they cannot be combined with that of HR (class D) (382). These studies have, incidentally, supplied supporting evidence that the TMV protein at neutrality or at slightly alkaline pH is largely trimeric.

The mechanism of the aggregation of TMV protein has been studied in some detail. It appears that below 12° the protein alone tends to aggregate in the form of stacked disks containing 17 subunits, rather than as the typical helix of 16 1/3 subunits per turn. At higher temperatures the formation of the helix is favored. The latter reaction is accompanied by the binding of one proton per subunit, probably a carboxylate group entering into hydrogen bonding in the $-C{\overset{\diagup OH}{=O}}$ form. The same has been found to occur on reconstitution of TMV (protein aggregation in the presence of RNA), and the obverse, the dissociation of about 2000 protons per virus particle, occurs on degradation of the virus (116, 121, 59, 260).

Reconstitution was used to advantage to prove that the TMV protein did not play a contributory role in transmitting viral genetic information, at a time when rigorous proof of this fact was still necessary. The experiment consisted in studying the progeny of mixed-reconstituted virus of the type discussed above. TMV RNA, suspected of retaining one or two TMV protein subunits, was reconstituted with HR protein, and vice-versa. The infectivity of such mixed viruses was inhibited by the antiserum against the added coat protein, proving that the mixed particles were really infective. However, the type of disease symptoms produced by such mixed-reconstituted virus was always that typical of the strain which had

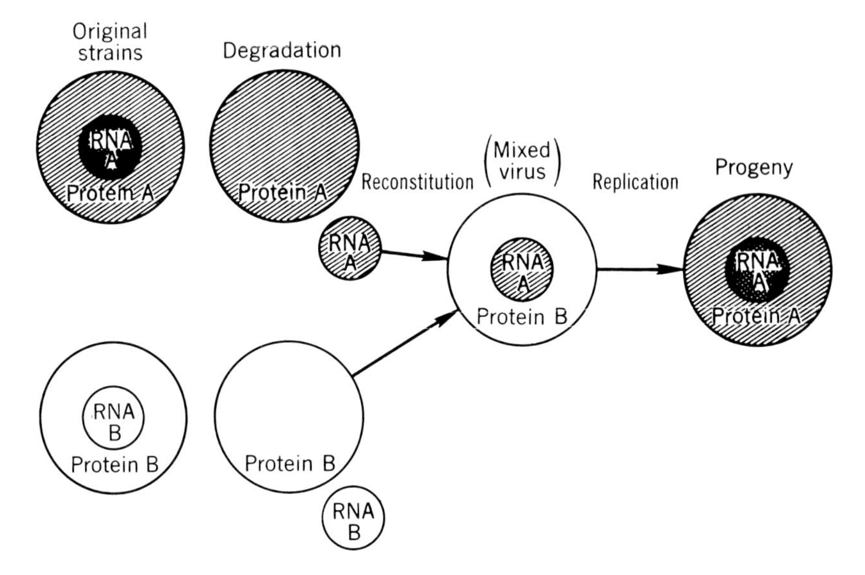

FIG. 10-2. Mixed reconstitution. Mixed virus is reconstituted from RNA of strain A and protein of strain B. The progeny of such mixed virus has all the properties, including amino acid sequence, of strain A (125).

supplied the RNA, and the progeny virus isolated had the protein composition of the strain which had supplied the RNA, even though it was reconstituted with the protein of the other strain. The genetic competence of viral RNA was thus established beyond doubt (Fig. 10-2) (125).

C. Reconstitution of Other Viruses

Only in recent years has some success been reported in the reconstitution of isometric viruses. Several small plant viruses (BMV, BBMV, CCMV) are dissociated by raising the salt concentration to about 1 *M*, and these proteins reaggregate to particles of viruslike appearance in the presence of nucleic acid on lowering the salt concentration by dialysis (24, 186, 495, 25). Thus the reconstitution conditions are quite unlike those required for TMV even to the point of not requiring appreciable time or elevated temperatures (0° about 1 hr). Thiol compounds (e.g., dithiothreitol) however, must be present to protect the —SH groups of the protein.

Although the RNA of the viruses of this group usually shows two or three components, reconstitution of this RNA mixture with any of the corresponding proteins produces one greatly predominant type of particle containing the equivalent of one RNA molecule (10^6 daltons).

Surprisingly, the proteins of these viruses not only form particles of viral appearance with the homologous RNA, or with RNA of other viruses of this group, but also with quite unrelated nucleic acids such as TMV RNA, ribosomal RNA, tRNA, and the RNA or DNA of small phages (186). Particles formed with TMV RNA show sedimentation values up to 152 S in which the infectivity of TMV RNA was enzyme resistant (551). The complex with ribosomal RNA shows a dual peak corresponding to the two components of ribosomal RNA. The fact, previously mentioned, that the virus particles of this group show dimensions and protein content similar to those of TYMV, yet contain only half as much RNA, makes it comprehensible that twice the normal amount of RNA can be incorporated. It is rather surprising that, on reconstitution, only the "normal" amount of the homologous RNA, but twice as much of a foreign RNA, can be accommodated. Another unexpected result of these experiments is the finding that the seemingly unrelated proteins of the viruses of this group are able to form particles with mixed coats (495). Viruslike particles are formed also in the absence of nucleic acid, although they, unlike other viral top components, differ from the virus in electrophoretic mobility (25).

The infectivity of the naked RNA of these viruses is similar to that of the undegraded virus, as well as of the reassembled virus, and thus increased infectivity cannot be used as criterion for successful reconstitution. However, the intraviral RNA is resistant, and the naked RNA sensitive, to nucleases and phosphodiesterase, and the protection afforded by the particle assembled *in vitro* is indicative of successful reconstitution of the original particle conformation. Yet one difference was noted: reassembled BMV was insoluble at pH 5, whereas natural BMV is soluble at this pH.*

For one other, slightly smaller class of viruses, the RNA phages of the f2 group, conditions have recently been developed which lead to the aggregation of viral RNA and viral protein to viruslike particles. The particles formed from MS2 coat protein and MS2 RNA

* The reversible dissociation of cucumber mosaic and alfalfa mosaic virus were reported in 1969 (542, 536b).

under appropriate conditions appear identical with the original phage on electron micrographs, but they sediment with 69 S instead of 81 S, and they are not infective (453). Very similar results were obtained with the fr phage (194). The nature of these noninfective particles resembles that of certain genetically defective strains (amber mutants) which were identified as lacking the maturation factor, the histidine-containing protein now termed A protein, one molecule of which occurs in the wild type phages of this group (see Section VIII,C).

Successful reconstitution of some infective R17 was reported when a little of the A protein was added to the reconstitution mixture of coat protein and RNA, suggesting that complete virus had actually been produced *in vitro* (355). The low level of infectivity detected may be due to the insolubility and intractability of the A protein, which may not have renatured adequately to function properly. Its actual function is still unknown beyond that it is required for infectivity, and that it protects the intraviral RNA against nuclease attack, possibly by keeping it from extruding from the particles. It is tempting to envisage the A protein as a monomolecular phage tail.*

The role of the RNA in the assembly of f2 type phage particles is important but not specific, since nonphage nucleic acids of greater or lesser molecular weight are also effective. The assembly of fr phagelike particles has been observed to occur at pH 7.8 even in the absence of RNA (180) and, in the case of CCMV and BMV, physical identity in the electron microscopic appearance of the *in vitro* aggregated protein particles and the virus has been reported 112a).

The ability of many other virus proteins to aggregate *in vitro* has been studied, but no clear-cut evidence for the formation of typical particles has been presented. The recent demonstration that the empty capsids formed *in vivo* may represent a necessary stage in the development of poliovirus is of considerable interest (222, 223). Assembly of the components of the T-even phages is discussed later (Section XI,C).

* Similar success in the reconstitution of partially infective Qβ phage was recently reported, also requiring a second protein (540a) (See Section V,B).

CHAPTER XI

THE REPLICATION OF VIRUSES

The subject matter of this chapter and subsequent ones is too broad and deep to do justice to it here. Many important papers are quoted and not discussed; others equally important are not referred to. It is hoped that the reader who wants to learn the current state of knowledge anywhere within the area of these chapters will be able to achieve his objective by reading the cited papers, and particularly the review articles listed.

A. Preliminary Stages and Events

The replication of viruses is preceded by their entry into a host cell and the removal of their envelopes and/or capsid protein. Then the nucleic acid is transported to the sites where it is replicated, transcribed to RNA if need be, and translated. Finally, the progeny components mature to virus particles and usually leave the host cell through budding, extrusion, or lysis of the cell. Relatively little is known in a definitive manner about the early and late stages of this process.

1. *Host Specificity*

Many viruses have become adapted to the point of infecting only one particular host species, while other viruses, such as TMV, can attack a wide range of hosts. The most striking examples of viruses having a wide host range are those plant viruses which also replicate in the insects which transmit them, and many of the encephaloviruses and other animal viruses, which replicate in both vertebrate and arthropod cells. No viruses are definitely known to replicate in both procaryotic and eucaryotic cells (bacteria and plants or animals).

The host specificity is in part a property of the coat protein, since infectious RNA has been shown in several instances to be able to infect tissue not susceptible to the intact virion. This was most clearly demonstrated with the poliovirus, which is infective only for primate cells, whereas rabbit kidney cells could be infected with polio RNA (196). In contrast, the differences among many of the host range mutants used for genetical mapping reside in the host's protein synthetic apparatus (see Section IX,C).

2. *Mechanism of Infection*

The first stage of attachment of viruses to host cells is not temperature-dependent, and it most probably involves ionized groups on both surfaces. Animal cells adsorb various picorna viruses at specific and different sites, and do not adsorb viruses to which they are not susceptible (195, 337, 73). Cellular virus receptor substances which are able to abolish viral infectivity rapidly have been isolated and have been characterized as glycolipoproteins. In some instances the purified receptor substance was shown to be able to release the viral RNA *in vitro* (338).

The small RNA phages become attached to the so-called F pili (about 4 per male bacterium), and the RNA subsequently penetrates via the pili into the cell (447–449). As with the animal cells, adsorption occurs in the cold, but penetration requires energy and elevated temperatures (37°). The fd (DNA) phages also infect only male bacteria, attaching themselves to the tip of the pili. These two types of attachment are illustrated in excellent electron micrographs (Fig. 11-1). According to some authors, the phage injects its DNA into the pili (57), but others believe that the phage enters

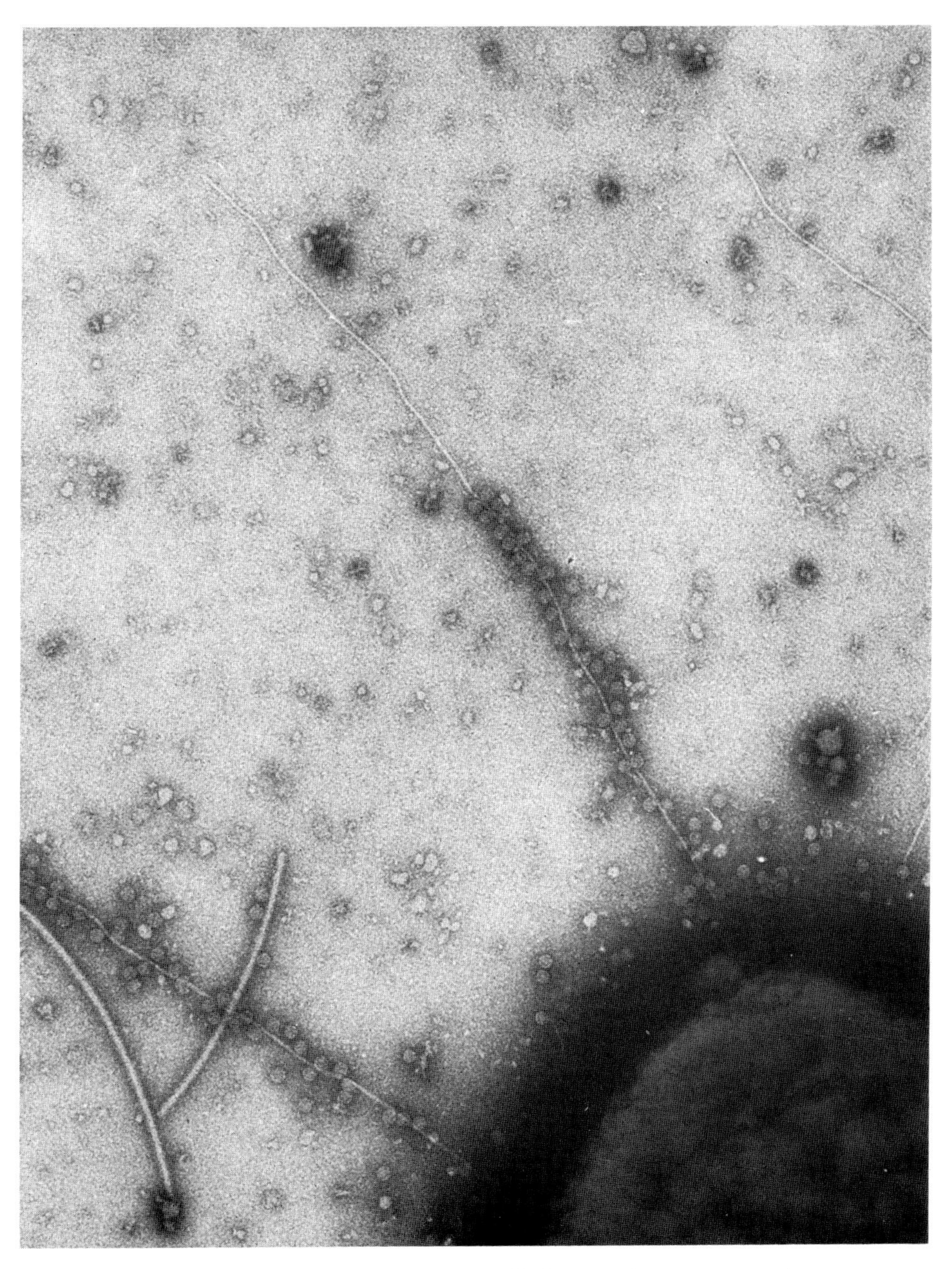

FIG. 11-1. Infection process with small bacteriophages. Two pili are shown emerging from an *E. coli* C600 cell, both carrying many attached MS2 (RNA) phages. A rod-shaped f1 (DNA) phage is also seen at the tip of one pili. × 63,000. (Courtesy of L. G. Caro.)

the cell intact (471). Since these are the only phages known to be released from the living cell, their intact entry may be by a similar mechanism.

ϕX174 infects both male and female bacteria, probably by a mechanism resembling that of the larger phages. That ϕX174 contains several capsid proteins may be a consequence of this more complex infection mechanism, and the pentameric spikes may represent the most likely attachment organ (425).

The myxoviruses and paramyxoviruses become bonded to receptors on the cell surface which are believed to contain sialic acid, although it is not yet established whether the viral neuraminidase serves for either entry or exit (see Chapter VII). In any case the virus particles can be seen to lose their envelope as they pass through the cell membrane, a process that has been termed engulfment. Again adsorption, but not penetration, occurs in the cold (415). Little is known about the infection process of the encephaloviruses, except that it shows similarities to that with the myxoviruses.

Reoviruses also enter the cell by a similar process called viropexis (104), the intact particles becoming engulfed and appearing in intracellular vacuoles where they can be observed as they become degraded (77). Similar mechanisms appear to be operative on vaccinia and herpes infection. The viral envelopes are thus generally removed either in passage through the cell membrane or in phagocytic vacuoles, and only the nucleocapsids appear to occur in the cytoplastic "factories," the sites of viral replication (226). Herpes virus nucleocapsids and probably also those of adenoviruses are transported to the nucleus before complete uncoating (226, 244).

3. *Uncoating of Viruses*

The more complex and the firmer the packaging of the virus, the more time or effort are needed to achieve release of its nucleic acid. This is borne out by the eclipse period in animal viruses ranging from 2 hr for picorna viruses to 10–12 hr for the vaccinia and Rous sarcoma viruses.

On TMV infection several hours more are needed to obtain first symptoms than if the naked RNA is used as inoculum (130). Also, the study of the sensitivity to UV light of infected centers on the leaf surface indicates that viral RNA begins to replicate more rapidly on inoculation with RNA than with virus (412, 511). Considerable differences have been observed in this lag period for different TMV strains but not for their nucleic acids. Thus the more stable and more UV resistant wild type produces infective centers which remain sensitive to UV irradiation for a much longer time (6 hr) than those of U2 (2 hr), or those produced by inoculation with the RNA of either strain (1 hr). Somewhat in conflict with these data is the finding that about 25% of the ^{14}C-labeled coat protein is stripped rapidly from inoculated virus particles (406).

The uncoating process has been studied in more detail with the pox viruses than with others (226, 244). It appears that a first stage of release of all phospholipid and much of the protein starts immediately on infection of HeLa cells. At the end of this process the viral cores are still intact, and the DNA is still protected from DNase. After a lag of up to 1 hr, the length of which depends on the multiplicity of infection, the second stage of uncoating takes place, transferring up to 70% of the viral DNA to a condition in which it is susceptible to DNase added to cell extracts. Yet, intracellularly it is not degraded for several hours.

These studies suggest a complex mechanism, and it was shown through the use of inhibitors of protein synthesis that an uncoating protein must be synthesized during that lag phase. The implications of this in terms of messenger RNA release by a genome that has not yet shed its protein coat is discussed further in the next section.

B. Replication, Transcription, and Translation of Viral Nucleic Acid

1. *General*

Understanding of the general processes and mechanism of intracellular nucleic acid and protein synthesis has advanced greatly in recent years, although many areas of uncertainty and doubt remain. Since viral replication occurs by the same principal methods and paths, it might be most easily described after a review of the current concepts of these normal cellular processes (256, 257).

DNA replication is achieved by DNA polymerases which connect the deoxynucleoside triphosphates (pppdA, pppdG, pppdT, and pppdC, commonly symbolized as dATP, dGTP, TTP, and dCTP) in proper order on the DNA template strand, releasing pyrophosphate as each bond is formed (see Section VI,B) (256, 321, 328). The exact mechanism of the replication of both strands of a double stranded DNA is still in doubt. It now appears that DNA replication proceeds in relatively short sections which are then interconnected by the action of DNA ligases-enzymes that establish 3′–5′ diester bonds between oligo- or polynucleotides held in position by a complementary strand (327a, 328).

The production of messenger(m)RNA results, by the same method, from condensation on the DNA template of ribonucleoside triphosphates through the action of DNA-dependent RNA polymerases. The mRNA then combines with ribosomal subunits, first binding the 30 S particles, to which the 50 S component adds later. In the presence of the many species of transfer(t) RNA, each charged with one of the 20 protein-building amino acids, a given triplet of nucleotides of the ribosome-bound mRNA combines with that tRNA which carries the complementary anticodon on its binding site (Fig. 11-2a,b). The ester-linked amino acid on that tRNA is then transferred to the amino group of the neighboring amino acid held in position in the same manner on the adjacent messenger triplet via the proper tRNA. The reading continues through movement of the ribosome along the mRNA (or vice-versa), and the peptide chain grows in the process.

The principal features common to all virus infections are: an eclipse or latent period during which little or no infectivity can be detected in the host cell; a period of increase in RNA and/or DNA (messenger and viral); a period of increase in proteins, the coat proteins gradually becoming predominant; a period of virus maturation; and a feature not common to all viruses: a period of release of virus from the cell due to budding and extrusion, or associated with cell lysis, and death. However, it appears probable that virus in nonmatured form (possibly naked nucleic acid) can also pass from cell to cell in the infected tissues, using organelles such as the plasmodesmata in plants. These five phases generally have a characteristic length for a given virus-host system, although the eclipse period is quite variable when a single animal virus particle infects a cell.

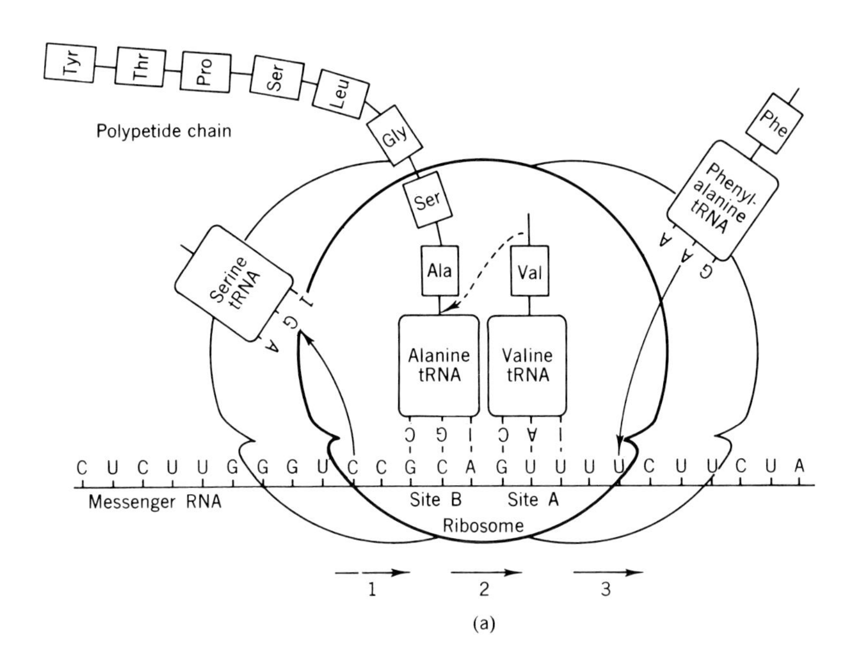

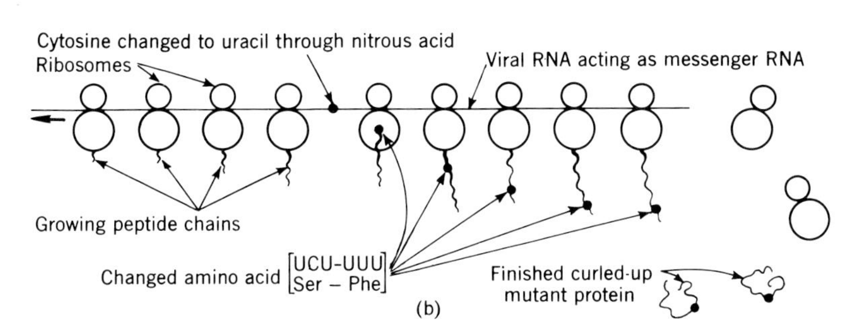

FIG. 11-2. Translation of messenger RNA nucleotide sequences into amino acid sequences. (*a*) The messenger (m) RNA, transcribed from the DNA, passes through a ribosome. Each coding triplet encounters and binds the complementary anticodon of an amino acid-charged transfer (t) RNA. I (inosine), occurring in the anticodons of many sRNA's, can bind to C, A, or U as illustrated. The ribosome appears to form two binding sites for tRNA molecules on the mRNA, one for positioning a newly arrived tRNA (site A) and another for holding the growing peptide chain (site B). (Courtesy of F. H. C. Crick.) (*b*) Translation of mutational event (C → U) to mutant protein (Ser → Phe). In both (*a*) and (*b*) the ribosomes move to the right (or the mRNA to the left).

The mode of action of interferon, an antiviral protein produced in virus-infected animal cells, has received considerable attention in recent years (218, 186a). It appears that the action of interferon to inhibit virus replication in animal cells is expressed at the level of translation of the viral (messenger) RNA (466). Thus it becomes comprehensible that interferon action is not virus-specific, not even to the point of distinguishing between DNA- and RNA-containing viruses. Recent studies of the requirements for the induction of the synthesis of interferon, or rather its precursor protein, have shown that not only viruses, viable or inactivated, but also double stranded polyribonucleotides are highly effective (186a, 67).

2. *Animal DNA Viruses*

The components of the large and medium large animal DNA viruses seem to be synthesized in the host cell in a manner very similar to the normal processes (244, 228, 227). The two paths are DNA replication on the one hand, and translation to messenger RNA followed by transcription to proteins on the other. Replication probably proceeds in toto (from one end of the DNA molecule to the other). The required DNA polymerase and ligase, although similar to those of the host, are probably virus-specific in the case of most animal virus infections. Thus the other functions of the viral DNA, its transcription and translation, must be initiated before replication if virus-specific enzymes are required for replication. These processes must, at least initially, utilize the host's enzyme machinery. The translation process is regulated by various control mechanisms so that certain parts of the DNA molecule are translated first and others later. Thus, early, middle, and late functions (i.e., mRNA's and proteins) have been differentiated in all virus-replicating systems. Many of these products have not been identified in regard to function, but shutoff mechanisms for the synthesis of host DNA and host protein are surely among the early functions.

In the particular instance of the replication of the *vaccinia virus*, the first event, preceding the release of the viral genome, seems to be the appearance of an uncoating protein. It is not yet clear whether this protein is coded by the viral or the host DNA. It seems possible that the necessary information is transcribed by a polymerase present in the virus core (244, 228). Several aspects of pox virus infection may become clarified by an understanding of the uncoating mechanism; among them are the phenomena known as pox virus initiation and reactivation, as well as the cytoplasmic "factories" in number proportional to the multiplicity of infection, where virus is being replicated and matured. Among the early enzymes made after uncoating are thymidine kinase, DNA polymerase, and various nucleases.

Structural viral proteins detected by serological methods or electrophoresis begin to appear after 2 hr, and up to 20 components can be seen after 20 hr, the end of the slow replication cycle of this virus. On the other hand, the rate of appearance of viral DNA is maximal after 2.5 hr and insignificant after 5 hr. The first complete virus appears after 6 hr, but two intermediate stages of coating with different capsid components can be detected during the preceding 2-hr period (16, 199).

The control of the transcription of a very large viral DNA such as that of vaccinia, which codes for as many as 500 proteins, is obviously very complex. Not only are viral functions initiated and host functions stopped, but also some of the early viral functions are found to be "switched off" in some positive manner at a later stage (300, 324, 301).

In contrast to the pox viruses which replicate in cytoplasmic factories, the herpes and adenoviruses appear to be replicated in the nucleus. The mode of replication of the herpes virus is illustrated schematically in Fig. 11-3.*

3. *Small RNA Viruses (Picorna Viruses of Animals, Plants, and Bacteria)*

The double stranded DNA viruses have been chosen as the first example of virus replication because their replication mechanisms most nearly resemble those occurring in the normal cell. The replication of the simplest viruses, containing linear RNA which can be

* Fenner's book (*The Biology of Animal Viruses*) (109) summarizes present knowledge of the replication of the various types of animal viruses in a similar schematic manner. A new important finding is the appearance of double stranded RNA, high in interferon activity (see p. 188), in vaccinia-infected cells (556). The existence of a DNA-dependent RNA polymerase in this virus is also getting more firmly established (228 and 557).

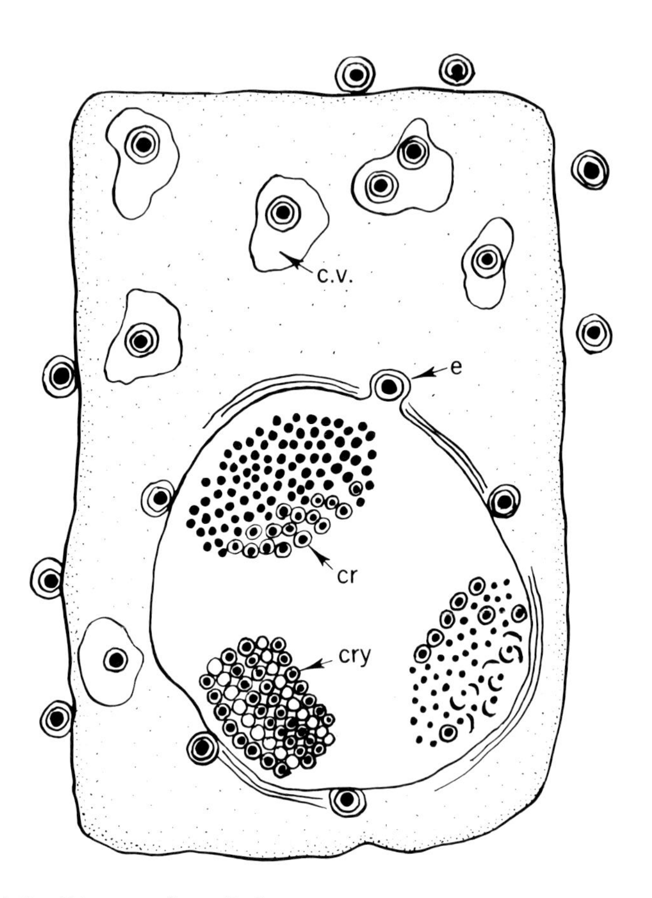

FIG. 11-3. Diagram of a cell showing herpes virus replication. Viral DNA and the icosahedral capsids (cr) form in the nucleus, and sometimes the capsids are arranged in a crystalline array (cry). Capsids acquire an envelope (e) by budding through the nuclear membrane and are then found in cytoplasmic vacuoles (c.v.) from which they are eventually released by egestion (109). (Courtesy of F. Fenner.)

regarded as analogous to the messenger RNA of the normal cell is now discussed, since two thirds of the processes of the replication of these viruses may be regarded as equivalent to those occurring in the healthy cell: nucleic acid replication and translation. That RNA $\leftrightarrow$ DNA transcription actually plays no role in the replication of the small RNA viruses of plants, bacteria, and animals has now been firmly established. On the other hand, the fact that the genome that must be replicated is RNA rather than DNA has represented a new challenge, both to the evolutionary history of viruses and to the ingenuity of experimental biochemists.

Extensive studies have been performed on the replication of several of the animal picorna viruses (polio, encephalomyocarditis, foot and mouth disease viruses), as well as of the f2 class of bacteriophages. The study of the replication of the plant viruses has not advanced as far, owing to technical problems in working with plant tissue and obtaining uniformly infected material (381a, 39a). However, there is no reason to assume that plant virus replication differs in principle from that of similar animal and bacterial viruses.

The main events of virus infection (see Section XI,A) have been found to occur with picorna viruses about 6 times faster in bacteria than in animal cells (1 hr vs 6 hr for the complete cycle), and the rate is even slower in plant cells (see Fig. 3-2).

Among the quite early effects of infection in animal cells is the *arrest of host RNA and protein synthesis*, but this is not so clear-cut in bacteria (189). Also, different hepatoma cell lines respond differently to Mengo virus infection, one showing the shutoff of cellular functions and the other not, while virus growth is similar in both (340). For this reason and others, it is believed that the shutoff of host syntheses is a coincidental result of a not yet understood aspect of early viral development. Mechanisms are required to enable viral RNA to change over from messenger to template function, and it has been suggested that these are the primary purposes of the early products which also arrest host biosyntheses.*

Another early, and indubitably the most important, translational event is the synthesis of an RNA polymerase, now usually termed replicase. In the case of the phage Qβ, where this enzyme has been intensively studied, it appears that it is a complex of two activities, carried by a single protein. The function of the replicase is bivalent dual in that it first utilizes the parental viral RNA, termed the "plus" (+) strand, as template in producing a complementary

* Evidence that arrest of host biosyntheses is not an intrinsic activity of R17 phage infection is given in a recent paper (555).

"minus" (—) strand, and then makes complementary plus strands (progeny RNA) on this template. Evidence has been presented that purified replicase contains macromolecular factors at least one of which occurs also in uninfected cells (135, 402). These factors seem to be required for the first of these steps (536a).

The difficulties inherent in isolating such enzymes free of nucleases which degrade the product were first overcome by Haruna and Spiegelmann in 1965 (170, 171, 428). It then became possible for the first time to observe replication of viral RNA *in vitro,* and to isolate progeny RNA which showed the same physical and chemical properties as the parental RNA added as template. But most important was the demonstration that the newly synthesized RNA was infective (Fig. 11-4). It now appears that neither the negative strand nor the double stranded RF and RI (see later) are infective *per se.* However, the negative strand is able to serve as template for the *in vitro* synthesis of Qβ RNA with the purified replicase, causing the appearance of infective plus strands without the lag period observed with viral (+) RNA as template (106a, 310). These properties of the negative strand are to be expected: it is not active as a messenger (being the complementary chain); and thus it cannot elicit the synthesis of replicase *in vivo* (nor *in vitro*). When supplied with that enzyme *in vitro,* however, the negative strand represents the direct template precursor for production of plus strands. As a result of these studies on Qβ, the demonstration of *in vitro* biosynthesis of any viral nucleic acid now appears to be only a technical problem.

An important aspect of this work is the finding that Qβ replicase is quite specific in terms of acting only on Qβ RNA (and Qβ-derived fragments carrying the 5′- (or 3′-linked) end of that RNA), but not on MS2 RNA, and vice-versa. It must be recalled, however, that these two phages are not closely related, since Qβ is not a member of the f2 group. An interesting extension of this work is the replication of Qβ RNA under conditions of stress which results in "evolutionary changes." When the time allowed for replication was shortened progressively to 5 min between "generations" of RNA, the product became gradually shorter, and became replicated faster (up to 1/6 and 15-fold, respectively), while retaining the same 5′-end group (pppG-, see Table 6-1) (309).

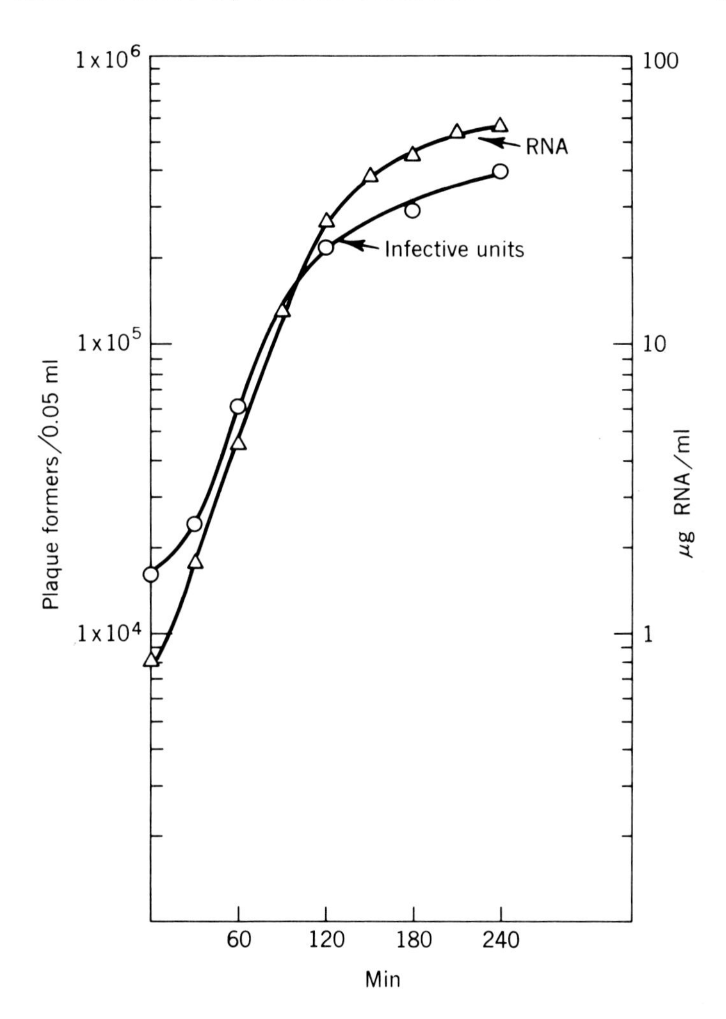

FIG. 11-4. *In vitro* replication of infective Qβ RNA. The reaction mixture contains the four nucleoside triphosphates, about 1 μg of Qβ RNA as template, and Qβ replicase (428).

The exact mechanism of action of the replicases of the small RNA viruses is still in some dispute. It would be envisaged *a priori* that each act of replication leads to a fully double stranded molecule; such double stranded forms have been detected in RNA-infected cells, and they have been termed the replicative form (RF).

In addition, there is ample evidence for the existence of molecules in which the RNA is only partly double stranded, replicative intermediates (RI), and schemes were suggested to account for them by postulating the simultaneous production of multiple complementary copies which gradually displaced one another from the single template molecule (189). Figure 11-5 illustrates this

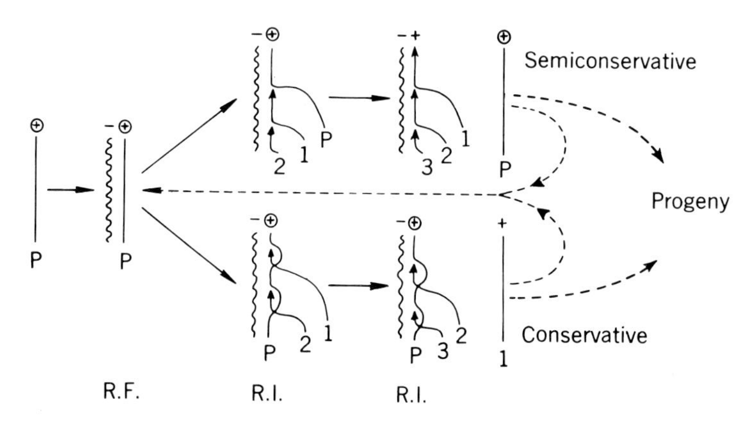

FIG. 11-5. Mechanism of replication of viral RNA. The nature of RF and RI, the replicative form and replicative intermediate, is illustrated. The latter allows the simultaneous synthesis of several (probably six) plus strands on one minus strand. The principle of conservative and semiconservative replication is also illustrated (189).

mechanism for both the conservative and the more probable semiconservative method of replication, which are also illustrated in the figure.

On the other hand, it has been shown that the act of extraction and deproteinization of viral nucleic acids can favor the annealing of actually separate complementary molecules, so that double strandedness can be an artifact of the phenol and the detergent methods of isolating intracellular RNA. Evidence has been presented favoring the concept that a complex of a template minus strand with nascent plus strands is held together not primarily by base pairing, but in some looser manner, possibly by the replicase molecules (106, 508). The direct capability of the nascent RNA to act as messenger and bind ribosomes may be a contributing factor to the tendency to remain or to become detached quickly from the template. Yet other recent studies have supplied new evidence for the existence of the RF and RI forms of the RNA of phages, as well as of plant viruses (39a, 204, 23). The recent analysis for nascent 5′-end groups (pppG-) of f2 RNA in RI type material also supports the concept of multiple simultaneous replications of minus strands, starting at the 3′-end (356). Final conclusions concerning the mechanism of viral RNA replication will have to await future experimental results.

The discussion of RNA genomes which are translated early led to consideration of the function of the most important early products, the replication of the RNA molecule which is simultaneously template and messenger. In the infected cell this dual nature also appears to be expressed simultaneously, although the most important late messenger function, coat protein synthesis, probably occurs after a considerable build-up of new viral RNA (+) strands.

Attempts to study the *biosynthesis of viral proteins* in *in vitro* systems have met with only partial success. When a stable *E. coli* cell-free system for the study of amino acid incorporation was developed, viral RNA was found to be an effective messenger in quantitative respects (322). The first RNA's to be used were TMV RNA and f2 RNA, and the first protein products to be sought were the coat proteins of these viruses. As far as the TMV system goes, no sound evidence for *in vitro* synthesis of that coat protein was detected (2), and the peptide pattern obtained on tryptic digestion of the biosynthesized products did not closely resemble that of the TMV peptides (396). Since it had become established in the meantime that all proteins of *E. coli* are synthesized with *N*-formylmethionine as initiator of the peptide chain, the effect of formyl donors and of formylmethionine on *in vitro* TMV protein synthesis was investigated, with no more success in terms of product identification (396).*

In contrast, with the RNA of the coliphage f2 as messenger, the tryptic peptide pattern of the proteinaceous products resembled that of the phage coat protein (318). Yet the viral coat proteins synthesized *in vivo* and *in vitro* differ in that only the latter carries the formylmethionyl end group (which is typical of all *E. coli* pro-

* However, it was recently shown (548) that TMV protein was synthesized and incorporated into virus rods in an *in vitro* system utilizing tobacco chloroplasts.

teins, and apparently split off in the cell before the protein becomes functional) (4), and there is also a deficiency in the amount of C-terminal peptide (residues 107–129) formed *in vitro* (270a). Translation of MS2 RNA appears to proceed properly but either imperfectly or incompletely under the conditions now used.

Polyacrylamide gel electrophoresis in SDS has proved to be a technique which gives more positive and very useful data on the translation of phage RNA to proteins, probably because this technique conveniently overlooks protein-chemical detail. Three definite virus-specific protein products are resolved by this technique when it is applied to extracts from *E. coli* spheroplasts infected with MS2 and thus *in vivo* (Fig. 11-6) (489). The main and fastest moving component is identified as the coat protein by its lack of histidine and other features. The other two are the maturation factor and the very slightly moving RNA replicase, identified largely on the basis of their decrease or absence when mutants deficient in one or the other function are analyzed by this technique (19, 159, 56). When the products of *in vitro* translated MS2 RNA are fractionated in this manner, the coat protein and replicase peaks are present, but, unaccountably, not the maturation factor. A new method to detect biosynthesized proteins giving different terminal ^{35}S-labeled (formyl-)methionyl peptides has revealed the presence of all three (273). It has also become possible to study by these methods the control mechanism of translation for such polycistronic messengers, as discussed below (Fig. 11-7) (19, 159, 56, 273).

The obvious question arises, "Why are phage proteins, but not plant virus proteins recognizable after *in vitro* biosynthesis?" One possibility would be the preferred and predominant usage of different codons in plants and bacteria. This does not question the now well-established universality of the code (the term "universal" referring in this instance only to the life processes on one medium-sized planet of this solar system).

However, there have been reports, not yet substantiated in other laboratories, that the RNA of two plant viruses led to protein products in the *E. coli* system which showed peptide patterns similar to those of the viral coat proteins. These RNA's were derived from the tobacco necrosis satellite virus (Section VIII,F) (65) and the top a component of alfalfa mosaic virus (Section VIII,E) (481). They have molecular weights of only about 0.4 and 0.5×10^6, and they prob-

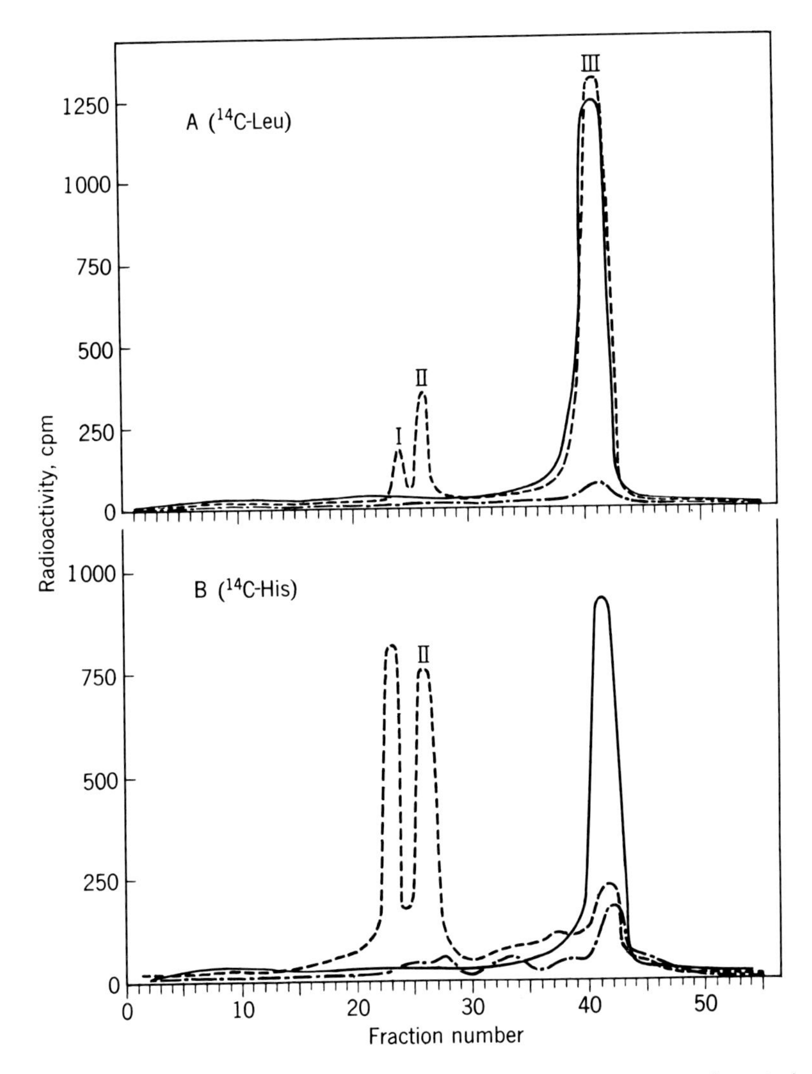

FIG. 11-6. Translation of MS2 RNA to three protein products (*in vivo*). Electrophoretic patterns (polyacrylamide gel) of A: ^{14}C-leucine-labeled, and B: ^{14}C-histidine-labeled products: – – –. Controls: – . – . – . ^{3}H-labeled coat protein as markers: ——— (489).

ably represent mono- or bicistronic messengers. Thus, if the faithful translation of these nucleic acids were confirmed by more discriminating methods, the conclusion would be that the polycistronic

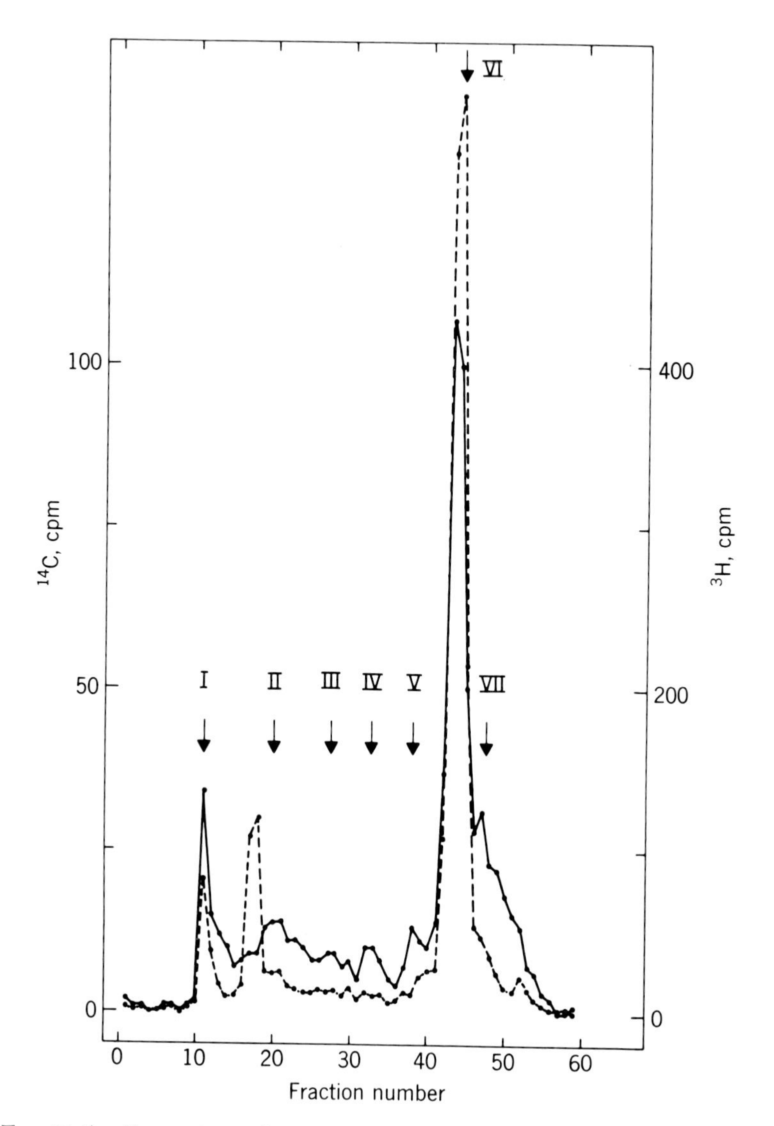

FIG. 11-7. Comparison of *in vivo* and *in vitro* products of translation of MS2 RNA. The dashed line indicates the three proteins made intracellularly (as separated on polyacrylamide gels). The solid line shows the more heterogeneous mixture of proteins resulting from *in vitro* translation of MS2 RNA, only the first (the replicase) and the third (the coat protein) being clearly present (452). Products II, III, IV, V, and VII are unidentified.

nature and size of TMV and TYMV RNA's, 2×10^6 daltons (only about 8% of which code for coat protein), interfere with the production and detection of the single now identifiable product, the coat protein, while the 1 million RNAs of the phages and the yet smaller RNAs of STNV and AMV-top-a represent less problematical test objects in this regard.

An aspect of greatest importance in viral replication, as it is also in normal cellular functions, is the *regulation and temporal control of gene expression* and protein synthesis. The RNA of the small phages which can code for maximally 1100 amino acids contains probably only the three cistrons so far mentioned: those describing the coat protein (129 amino acids), the maturation factor (A protein, about 350 amino acids), and the replicase(s) (possibly 450 amino acids). The control of the production of these three proteins is obviously important since a relatively small amount of the replicase must be synthesized early, while much of the coat protein and a small amount of the maturation factor are needed later.

That the coat proteins of the phages (and of TMV) contain no histidine, while the other two phage coded proteins contain this amino acid, has facilitated studies on their relative rates and amounts, and considerable progress has been made in the elucidation of this particular control mechanism. It appears that there is a marked tendency for one capsomere (5 or 6 subunits) of the MS2 coat protein to combine with the phage RNA, and that this complex is almost completely restricted in its messenger activity to direct the production of only the coat protein (451–453). Thus, at a time when RNA translation has proceeded to the point where approximately 6 molecules of coat protein and presumably of the other proteins have been synthesized per RNA molecule, the production of histidine containing proteins was greatly decreased and the lack of synthesis of replicase was evident from electrophoretic patterns. (Figs. 11-8, 11-9). By a new technique of detecting the three proteins coded by f2 RNA, evidence was obtained that only the synthesis of the replicase was depressed in the RNA-coat protein complex (273). The production of maturation factor, however, was at all times much lower than that of coat protein, although both of these proteins began to be made prior to the replicase. A specific complex of coat protein with the RI form of phage RNA has also been de-

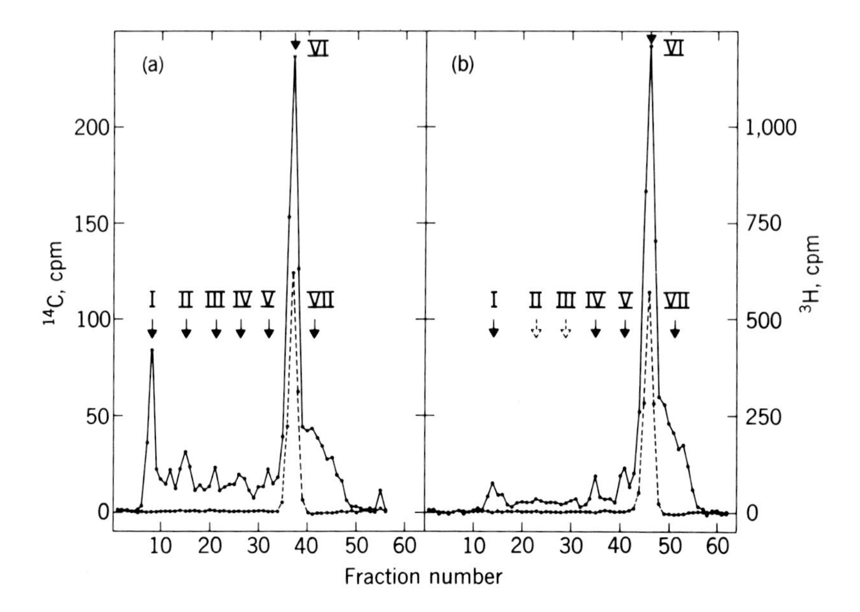

FIG. 11-8. Regulation of translation of viral RNA. The proteins produced by free MS2 RNA (*a*), as compared to complex I (MS2 RNA + MS2 coat protein subunits, 1:6 molar ratio (solid lines) (*b*), and added coat protein marker (broken line). It is evident that synthesis of all but the coat protein is greatly depressed by complex formation of viral RNA and coat protein (451–453).

scribed, and it has been suggested that the change in the relative levels of replication and translation play the major regulatory role (357). However, the latest experiments of Sugiyama combining *in vivo* and *in vitro* studies, favor the concept that the coat protein controls translation directly. The mechanism is not understood which enables a protein to interact with a specific site on an RNA, a phenomenon formally similar to the interaction of λc_I repressor and λ DNA (see Chapter XII). This represents a challenging area for future research.

It should be stressed that this interaction of one capsomere with the viral RNA is quite specific, and that MS2 protein does not interact with TMV RNA and vice-versa. The previously described reconstitution of phage particles (Chapter X) occurs under the same conditions when the ratio of proteins to nucleic acid is considerably higher. Thus phage maturation, the intracellular equivalent of re-

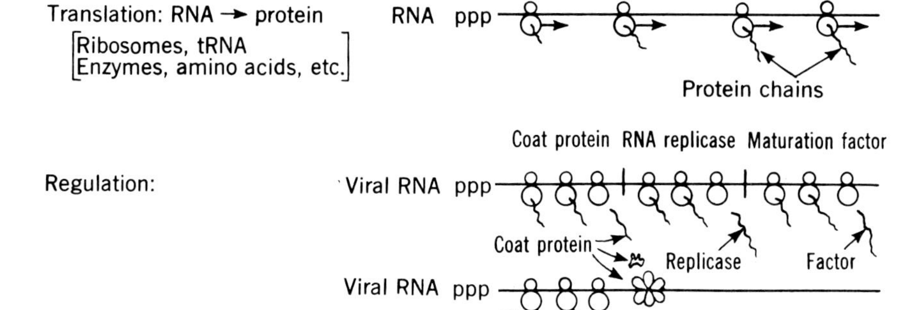

FIG. 11-9. Phage coat protein repressing reading of two cistrons. Schematic illustration of translation of a polycistonic viral (+) RNA strand.

constitution, may also be regarded as a control mechanism, at this stage removing the RNA completely from all its potential intracellular functions.

Control mechanisms involving a gene product have been discussed, but there are various indications that the conformation of viral nucleic acids can as such exert control over the translation of various segments (273). Thus a mutation of the RNA which expresses itself in the sixth amino acid from the N-terminus of the coat protein of f2 also greatly affects the production of replicase and progeny RNA (but not of the maturation factor), while mutations nearer the middle of the coat protein have no such "polarity" effect. Also, recent studies with the defective mutants previously mentioned have indicated that some of these mutants lack a good bit of RNA (14–18 S instead of 27 S), including the 5′-end (pppG-), and that this RNA is effectively, and without lag, translated *in vitro* into the histidine-containing proteins, replicase and maturation factor, but makes no coat protein (273).* In contrast, the intact RNA makes coat protein and maturation factor prior to the replicase. The various data obtained concerning *in vitro* trans-

* In contrast, the fluoro uracil mutants lacking the 3′ (right) third of the molecule, code for coat protein, but not for replicase in the cell-free system (408, 409).

lation of phage RNA's are not yet in clear accord, and the final word about the physical sequence of cistrons and the order and control of their translation cannot yet be said, although most of the data favor the location of the coat protein cistron at the 5′-end followed, probably, by the replicase (273).

In addition, it is by no means certain that in animal cells the same processes of multiple initiation and termination of peptide chains, read from a single polycistronic viral RNA, are operative as in bacterial cells. With regard to the multiple poliocapsid proteins, it has been clearly shown that they result from enzymatic cleavage of a large gene product protein, and it is quite possible that animal cells deal only with monocistronic messengers (222, 223).

Studies on the topography of picorna virus replication in animal cells indicate that transcription, translation, and maturation occur in the cytoplasm (72), although surely in association with membranous surfaces (334). In contrast, the RNA of several plant viruses appears to be replicated in the nucleus and predominantly in the nucleolus. Protein synthesis and virus assembly, however, occur in the cytoplasm, probably on the contact surfaces of membranes (61, 386).* Opinions differ concerning the role of the chloroplasts, if any, in virus synthesis. Electron micrographs of sections reveal the appearance of new and different structures of unknown nature on infection with various viruses ("noodles" in the case of TMV infection) (311).

4. *Viruses with DNA of Circular Tendencies*

The replication of the DNA molecules of unusual properties which characterize many of the phages and the small DNA viruses (groups 9 and 12 to 14 of Table 3.1) are now discussed.

The single stranded circular DNA of ϕX174 quickly becomes attached at particular intracellular sites and is transformed into a double stranded circle (RF) (see Fig. 11-10). A recurring facet of many phage infections, not stressed in the preceding discussion of RNA phage replication, and best studied in the case of ϕX174, is the singular role of the infecting parental strand, possibly only

* The site of synthesis of TMV protein and RNA continues to be disputed as evidenced by two 1969 reports (543, 546).

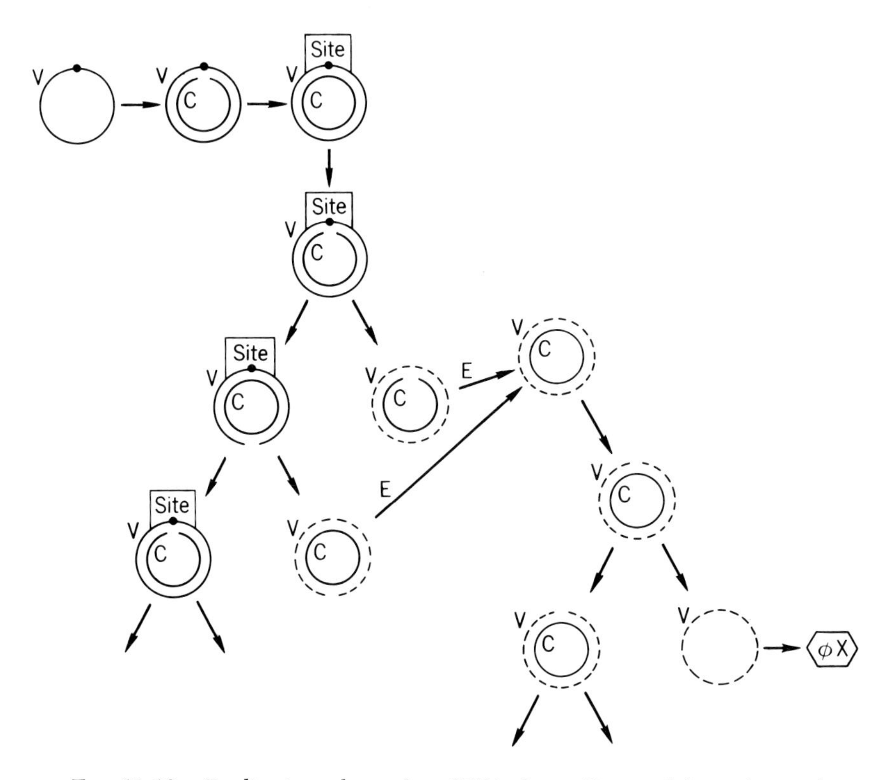

FIG. 11-10. Replication scheme for ϕX174 phage. V = viral (or +) strand, C = complementary (or −) strand (426).

one per cell, which does not seem to enter the pool of the progeny (+) strands, once they appear. The described replication scheme (425, 426) accounts well for this phenomenon, as far as ϕX174 is concerned. It appears that the double stranded closed circle formed from the invading DNA becomes attached at an intracellular site through the parental (viral) strand which remains there. The complementary (−) strand, however, becomes cut and released. Alternatively, the viral (+) strand becomes cut, and this leads to release of the intact circular complementary (−) strand. These released (−) strands, open or closed, serve as templates for production of (+) strands, which leads incidentally to repair of the template circle. Finally the new (+) strands are cut, released from double strandedness, again closed, and incorporated in the phage heads.

Recently it has become possible to observe the replication of ϕX174 DNA *in vitro* using highly purified DNA polymerase and polynucleotide joining enzyme (ligase) preparations from *E. coli* (153). To this end the parental single stranded circle was first used as template, and deoxybromouridine was used in lieu of thymine (triphosphate) for the purpose of producing a "heavy" (—) strand. This biosynthesized (—) strand was then separated by CsCl gradient centrifugation after "nicking" and denaturing. It was infective. This material was then used, in turn, as template. Infective (+) progeny could be obtained in this manner.

Much less is known about fd replication. It may be assumed that it resembles that of ϕX174. However, fd in contrast to ϕX174, does not lyse the cell but is extruded from it, and the resulting continuity of concurrent phage production and bacterial growth and division may well be assumed to affect the mode of replication of the phage.

While from a chemical viewpoint there is a great difference between a long linear and a large circular molecule, they are actually only one phosphodiester bond apart and biologically freely reversible, as is also the transition from single to double strandedness. This is even more true for the transition from the double stranded supercoiled to the double stranded open form, the latter usually resulting from a break in one of the chains (nick). That these easy transitions occur is borne out by the mechanism of replication of ϕX174. They are also evident from studies of the replication of the double stranded supercoiled DNA of viruses of the polyoma and papilloma group. Here also all forms that retain some circularity (supercoiled, open circle, and single stranded circle) are infectious; this means that they can initiate replication. Only the linear molecule is "dead" in this sense. The assumption is that replication in each of these cases requires the opening of a ring, but that the resultant linear molecule must be held in shape by its complementary unnicked partner, and that the final act of replication is the closing of the circle in the progeny strand. The appearance in phage-infected cells of new ligases or sealases which make diester bonds between 5′-phosphates and 3′-OH groups probably serves this purpose (30, 321).

Among the few facts known about polyoma replication is that it resembles other viral replications in being semiconservative, and that the double stranded progeny DNA is supercoiled as soon as it can be detected.

The situation with the λ phage is similar in that the double stranded state with cohesive ends, which is prerequisite to intracellular circularization, is also a prerequisite for infectivity. Thus both strands when separated after denaturation are noninfective (in the helper assay), but regain infectivity on renaturation (84). During the lytic response,* λ DNA is largely in the supercoiled circular form, which represents its *RF*. The ring is opened only at the time of virus maturation (530).

Besides the open viral form, all intracellular circular forms of λ are infective in spheroplasts, including the denatured, open circle form. This means that a single stranded ring suffices for infection (247, 248), in contrast to the helper assay, which shows strong infectivity only for the open double stranded viral DNA with intact cohesive ends.

There is the question: Why are certain phage DNA's always circular and others, as in λ and the T-even phages, are not? Possibly the relatively small amount of protein present in the particles of λ and the T-even phages (about as much as DNA) is not suitable for the packaging of circular DNA molecules; it is also plausible that a linear state of the DNA is of advantage for its insertion into the head and its ejection through the tail. In any case, only those phages which have tails and relatively little protein ($< 50\%$ of their mass) have linear DNA.

C. Replication of T-even Phages

Phage adsorption is affected by the ionic medium, but it often requires rather specific cofactors, the one most studied being L-tryptophan in the case of T4 (235, 439, 258). It appears that the binding of a few molecules of tryptophan by the phage gives a stable complex, and that the tryptophan affects the configuration

* See Chapter XIII for a discussion of lysogeny.

of the tail fibers, favoring their extension in the manner necessary for adsorption to the bacterial surface. Adsorption occurs at receptor sites which differ for different phages, being of lipoprotein nature for T2 and T6 and of lipopolysaccharide nature for T3, T4, and T7.

The process of infection, with T2 for example, requires the following steps: The attachment by the tail fibers is followed by that of the pins on the hexagonal base plate of the tail; then the tail tube is detached from the base plate; the tail sheath contracts and forces the tail tube through the soft layers containing the lipoproteins and lipopolysaccharides. The subsequent penetration of the rigid inner layer of the cell wall may involve the action of phage lysozyme (Figs. 11-11, 11-12) (416, 417).

The contraction of the tail sheath has resemblances to muscular contraction, although the role of ATP, if any, in this process has not been clarified. It seems that changes in the conformation of the base plate play a critical role in triggering its release from the tail tube and the subsequent contraction of the sheath (258).

The DNA with a diameter of about 2.4 mμ is then squeezed in some manner through the tail tube of 2.5–3.0 mμ inner diameter. Obviously one end of the 50,000 mμ long DNA molecule must be so positioned as to favor this movement, which is in any case difficult to conceive. But it unquestionably occurs, requiring about 1 min. The entire cycle from infection to burst of the host cell and release of 100–300 progeny phage particles requires only about 30 min.

The early functions on arrival of T4 DNA in the bacterium are the usual arrest of host synthetic functions. Instead, there starts a rapid production of phage DNA after about 10 min, as well as of unstable messenger RNA, complementary to the phage DNA and able to hybridize with it after heat denaturation (493, 15, 160, 268).

FIG. 11-11. Thin section electron micrograph of T4 phages adsorbed on *E. coli* B cell wall. The phages are bound to the bacterial surface by short fibers extending from the base plate. The base plates have what Simon and Anderson call a "boat type" profile in which the tips of the tail plate point away from the cell surface. The arrow indicates the tail tube of one of the phages which has penetrated through the cell wall. The visible thin fibrils of 3 mμ diameter probably consist of phage DNA (416, 417).

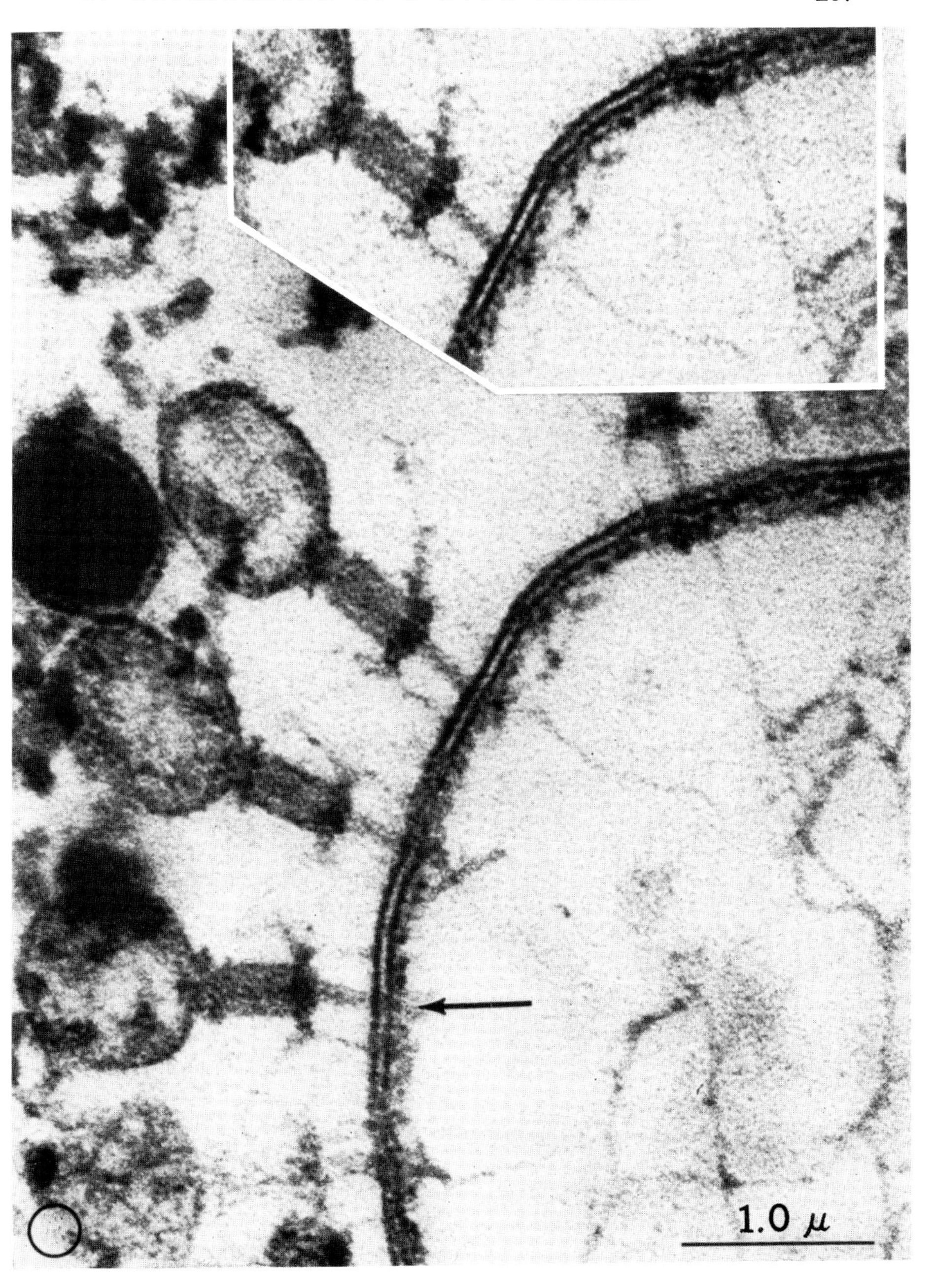
1.0 μ

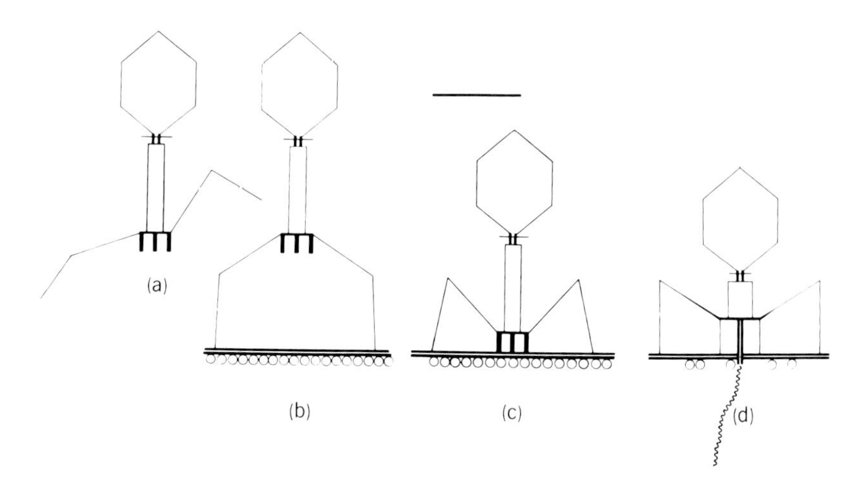

FIG. 11-12. The steps of attachment, penetration, and injection of the T4 and T2 bacteriophages to their host cell. (*a*) An unattached phage. (*b*) The long, thin tail fibers have attached to the cell wall. They are kinked slightly near their middle. The base plate of the phage is over 100 mμ from the cell wall. (*c*) The phage has moved closer to the cell wall, and the pins extending from its base plate are in contact with the wall. Long tail fibers are shown kinked and bent sharply at their hinge on the base plate. (*d*) The tail sheath has contracted, and the sheath has retracted up the tail tube, carrying with it the freed base plate. The phage is still linked to the host cell by short tail fibers. The inner tail tube has penetrated the outer three layers of the host cell wall. A decrease in the density of the rigid portion of the cell wall is also shown (416, 417).

Preceding these gross events, and within the first 8 min, a considerable amount of transcription and translation must occur to produce the many phage-specific enzymes that are needed before the synthesis of phage DNA can begin. Of greatest importance are the eight or more enzymes that are involved in the synthesis and incorporation of the new phage DNA component, deoxyhydroxymethylcytosine triphosphate, pppdHMC, and the arrest of production and removal of pppdC (Fig. 11-13). The synthesis of a new phage-specific DNA polymerase and ligase has also been demonstrated in the phage-infected cell (15, 160, 30, 321). Although *in vitro* the DNA polymerase of uninfected *E. coli* can use T-even DNA as template, the production of the phage-specific polymerase *in vivo* is an absolute requirement. Also, the glucosylating enzymes must be synthesized. The specific functions of a newly discovered

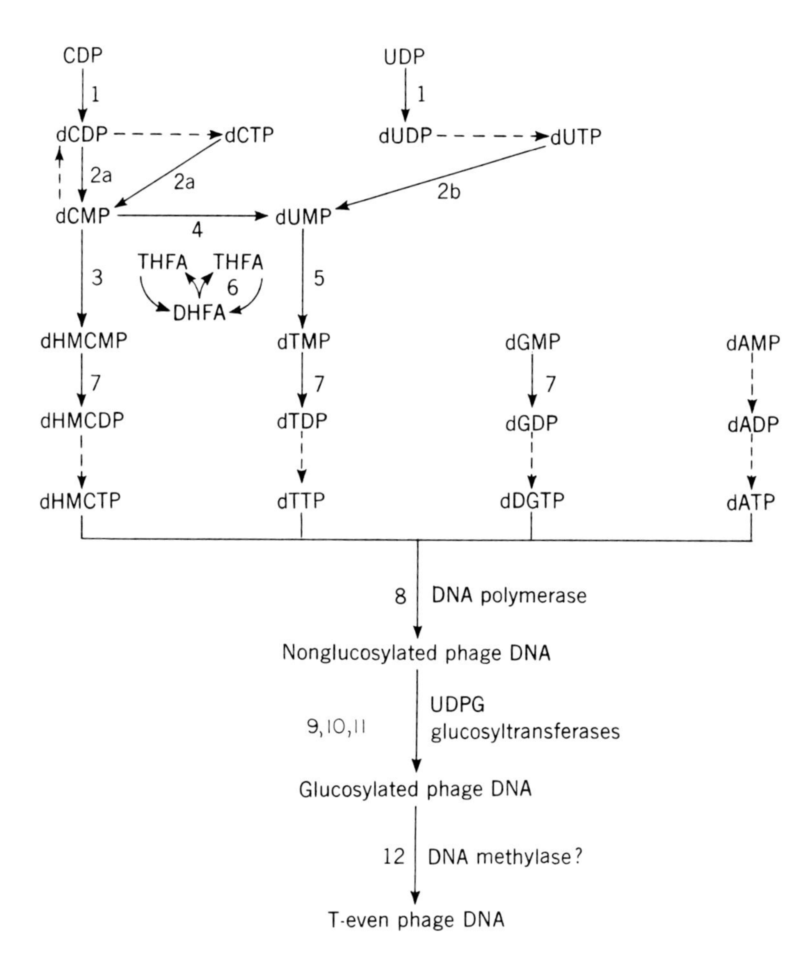

FIG. 11-13. Pathway of synthesis of T-even phage DNA. Solid lines represent reactions catalyzed by phage-induced enzymes; dashed lines, those catalyzed by host enzymes. The names of the enzymes are: 1, ribonucleoside diphosphate reductase; 2a, deoxycytidine di- and triphosphatase; 2b, deoxyuridine di- and triphosphatase; 3, deoxycytidylate hydroxymethylase; 4, deoxycytidylate deaminase; 5, thymidylate synthetase; 6, dihydrofolate reductase; 7, T-even phage kinase.

group of "very early proteins" have not yet been established (268).

When the phage DNA production reaches a certain level and appreciable amounts of phage DNA have accumulated, then, about 8 min after infection, synthesis of some of the early enzymes ceases, and synthesis of the capsid proteins starts. Another late product is the phage lysozyme. Complete virus particles begin to appear about 15 min after infection and, when they reach their maximum concentration, the cell bursts and the phages are released. The mechanisms of maturation have been studied extensively in recent years; they are discussed in the next section. Whether the phage lysozyme plays a role in this "lysis from within" and the bursting of the host cell has not yet been established.

The origin and fate of the phage components during infection are now quite well known. More than 50% of the material present in both parental DNA strands appears in the progeny, but no direct transfer of double stranded segments or full-length chains seems to occur (454). Most of the progeny DNA is derived from degraded host DNA, but *de novo* synthesis also takes place. As far as the proteins are concerned, a little of the internal protein may be reused, and some of the protein is derived from degraded cellular proteins, but most of the phage-specific proteins are newly synthesized (258).

D. Maturation and Release of Viruses

The maturation of viruses is best observed by electron microscopy, (see Fig. 3-2). The appearance of stable infective units in cell lysates, obtained at various time periods by cyanide or lysozyme, can also be used as a criterion for the completion of the process.

The studies on *in vitro* reconstitution of plant and small bacterial viruses indicate that the assembly of simple viruses is a spontaneous entropy-driven process which proceeds rapidly once the concentration of the protein and nucleic acid has become sufficient. On the other hand, the need for a maturation factor for the stabilization of bacterial picorna viruses has been discussed (Section IX,B), and this may well be a more general phenomenon, as suggested by results obtained with adenoviruses and herpes virus (376). As far as the maturation of RNA phages is concerned, when the progeny phage has become completed, usually about 10,000 particles per cell, the *E. coli* cell lyses, and the phage is released into the medium. No evidence for a lysing enzyme has been detected in these RNA phages, in contrast to the situation with the DNA phages.

The plant viruses remain in the cells unless they are transmitted to a fresh host by mechanical agents or biological vectors.

As far as the assembly of the large and medium-sized phages is concerned, extensive studies have been performed in recent years on the genetic basis of the formation of the various phage components. Two types of mutants of T4 and other phages have supplied the material for these studies: the temperature-sensitive mutants (ts), which propagate at 25° but not at 42°, and the host specificity mutants (am) (see Section IX,C), which propagate in *E. coli* CR 63 but not in *E. coli* B (99, 100, 246, 504, 505, 525, 219). Observation of the products resulting from infection under non-permissive conditions makes it evident that phage maturation can

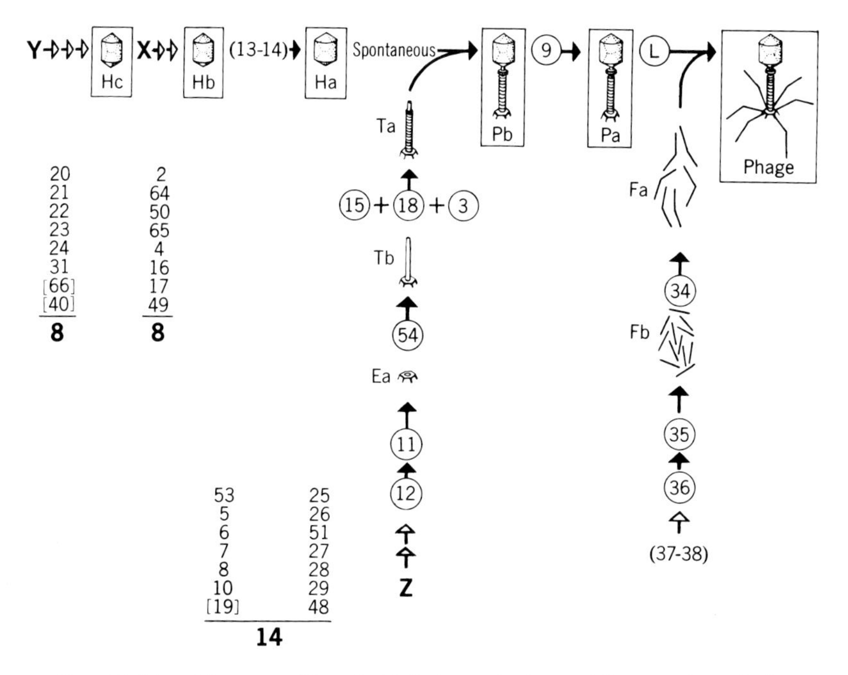

FIG. 11-14. The assembly of T4D bacteriophage. The numbers indicate the genes which have been identified as playing a role in the assembly of the phage particle. Two large sequences of genes, X and Y, act to produce a phage component such as the phage head. A number of genes are needed to produce a precursor of the tail plate, and they are called the Z sequence (99, 525). Total genes = 45.

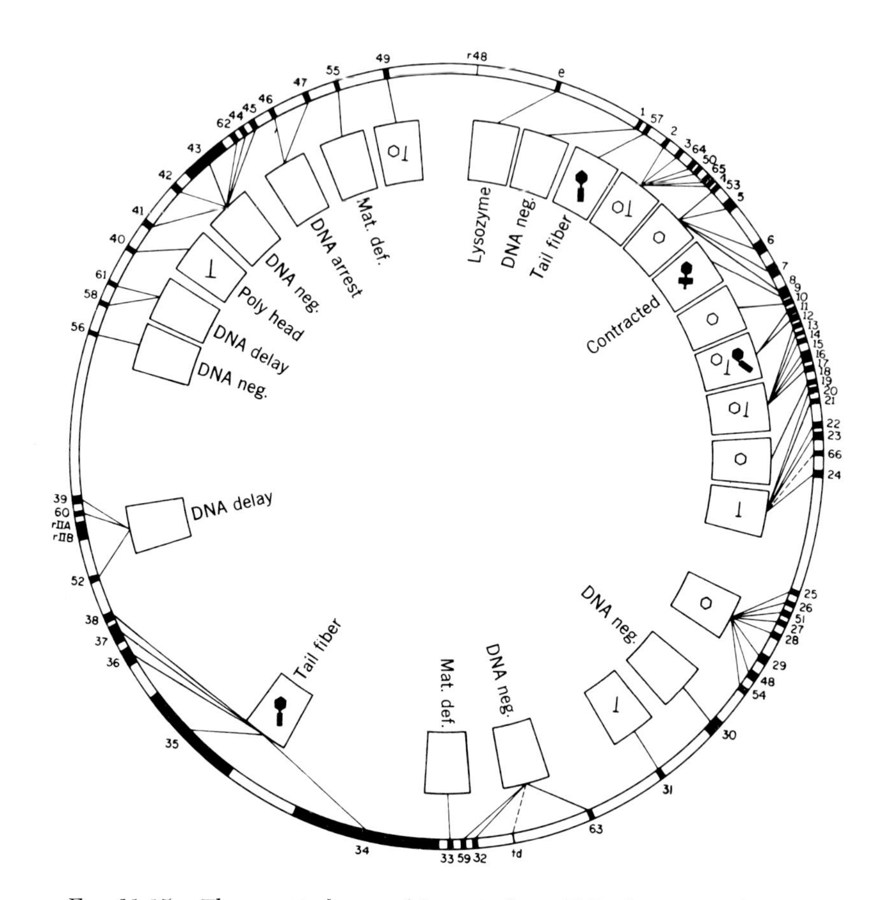

FIG. 11-15. The genetical map of bacteriophage T4D. Locations of cistrons defined by ts and am mutations. The effect of the mutation on phage formation is indicated in the central area (99, 525).

be blocked genetically in a variety of manners. It appears that at least 45 cistrons control the formation of a T4 phage particle, and that particles defective in one or another feature can now be produced at will (Figs. 11-14 to 11-16).

These investigations by Edgar and co-workers have made it pos-

---

FIG. 11-16. Crude cell lysate-containing defective phages. The defective phages SPα and SPβ are in center and at top (with twisted tail), respectively. Phage tails can be noted also. (Courtesy of F. Eiserling.)

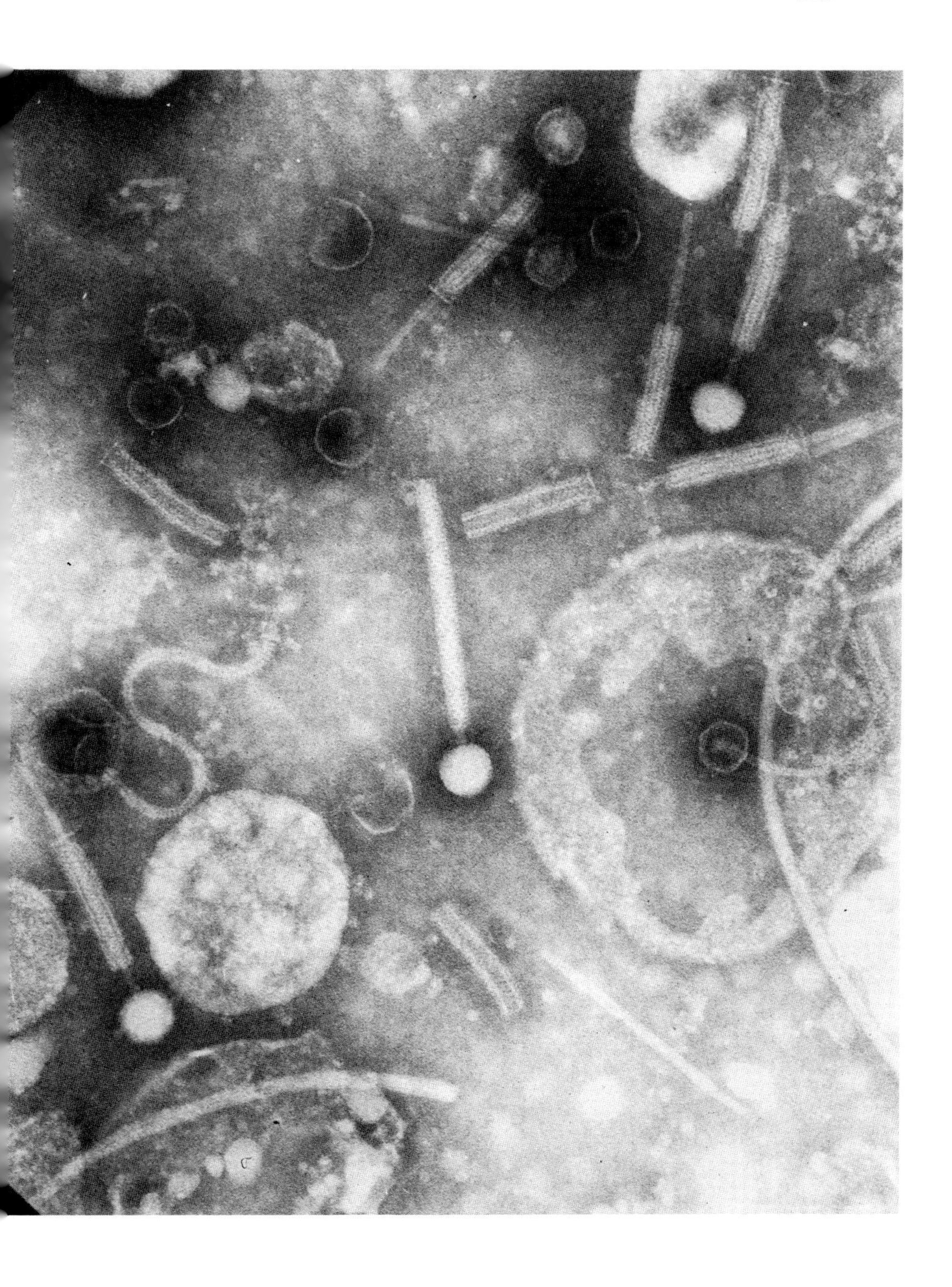

sible to study the assembly of complete particles *in vitro* by complementation of the variously deficient components. For example, the infectivity increases by 2–3 orders of magnitude if a tail-fiberless phage is combined with the tail fibers produced by another mutant.

It should be stressed that there is an important difference between the self-assembly of simple viruses and the maturation of complex phages, since many steps in that process require the action of specific phage coded enzymes. The combination of head and tail is apparently one of the few spontaneous steps, while most of the other steps take place *in vitro* but require extracts of the infected cell, rather than proceeding with the purified protein or nucleoprotein components.

One aspect of phage replication that is presently under active study is termed host-induced modification. It appears that certain bacterial strains (e.g., *E. coli* B/40) phenotypically modify the nascent phages in a manner which makes them unable to infect *E. coli* B but does not so affect their viability in other strains. This effect has been identified as a consequence of greatly decreased glucosylation of the phages (175, 406a). Another biochemical pathway of host-induced modification is alteration of the methylation patterns (536). These modified nucleic acids are susceptible to highly specific phage-induced nucleases.

CHAPTER XII

THE BIOLOGY OF THE TUMOR VIRUSES

A. General Remarks

The range of possible virus-host relationships can be compared to sociological events recorded in man's history. The typical virus disease corresponds to invasion of a civilized country by a nomadic enemy army which for a while lives off the land, takes what it wants, destroys the social order, and departs richer and more numerous, in search of another victim.

The nonapparent virus infections might then be likened to the passage of nondestructive outsiders through a country, such as the Russian émigrés who settled in Manchuria and northern China early in this century and then moved on to America.

Seen in such manner, viral oncogenesis resembles the role of the Romans in Gaul or the Normans in Britain, and similar occurrences in history. The invading army is more sophisticated, though still dependent on the host country's economy. The invader comes to stay, he enriches himself, but he also builds; and soon he becomes integrated into the existing society. He may gradually change the character of his new country, but in so doing he sacrifices some of his own identity. The "transformed" country may flourish, it may even become aggressive, but the individual of the invader's heritage is not moving on, invading, and despoiling new countries, for he has become settled in his new social situation which had proved acceptable, if not advantageous, to his offspring as well as to those of the native population.

From such a broad and generalized viewpoint, viral oncogenesis represents the mutual adaptation of virus and host cell, each undergoing genotypic transformations so that the new state may flourish. The key to such development is the integration of some genetic material and, thus, information from both partners.

There are tumor viruses among most of the major groups of DNA viruses, but the most studied and most frequent oncogenic viruses occur in the adenovirus and papova groups. With these DNA viruses the principle of viral and host genetic interaction and integration is not difficult to conceive.

Almost all oncogenic RNA viruses belong to one more or less closely interrelated group (for which the name leukoviruses has been proposed). Here integration requires RNA-DNA interactions and possibly reverse transcription: DNA synthesized on an RNA template. This presents interesting challenges to future biochemical research.

The problems of viral tumorogenesis have been intensively studied in recent years, stimulated by the hope that such studies will yield useful information in regard to malignancy in man. Obviously, the biological complexity of the phenomena has represented a strong intrinsic challenge to the ingenuity of the investigators. However, the facts that most oncogenic viruses never reach high concentrations, and that many of them lose their viral integrity in the course of oncogenesis has made them rather difficult objects of biochemical research. The preferred tool in these studies has been immunology, but more recently the availability of viruses labeled with high levels of radioactivity has opened up the field to the methods of molecular biology, such as sucrose gradient fractionation and hybridization studies.

B. The DNA-Containing Tumor Viruses

The most effective oncogenic DNA viruses fall into two groups which have some similarities and apparent interrelationships, yet many major differences. The polyoma and papilloma (papova) viruses are comparatively small (30–50 mμ diameter) and contain circular DNA (3 and 5 $\times 10^6$ molecular weight). The adenoviruses are two to three times larger, and their linear DNA has a molecular weight of about 22 $\times 10^6$ daltons. The existence of both the polyoma and adenoviruses was only recognized during the 1960s because they are generally latent and cause no apparent disease or very minor disturbances in the tissues of man and various animals.

The first mammalian tumor to be identified with a virus disease (in 1933) was the rabbit papilloma (410). The keratinized outer layers of these skin tumors in cottontail (but not in domestic) rabbits are rich in virus (364), and the virus is probably transmitted by arthopod bites. However, if these benign tumors become malignant, a not infrequent occurrence, then virus is no longer detectable in them. It has nevertheless become evident in recent years that infectious DNA can frequently be extracted from carcinomatous tissue resulting from papilloma virus infection (220). The transformation of rabbit cells or tissues is accompanied by a marked increase in arginase activity, but it is not certain whether this enzyme is virus-specific (331).

The study of viral oncogenesis, like that of so many biological phenomena, was greatly stimulated by the discovery of new and more reliable and rapid means of assaying the biological response. This was the finding that many newborn animals, and particularly rodents, are much more sensitive to oncogenic viruses than are the adult animals (209, 210). Hamsters are more responsive than rats or mice, even when adult (472). The general interpretation of this phenomenon is that in newborn animals the antibody-producing responses have not yet been developed. It has become evident that the presence of actual or potential antibodies represents one of the main means of an organism to combat the transformation of its cells into tumor cells.

Newborn animals proved particularly advantageous for use with the polyoma group of viruses. These viruses (440) show close similarities and interrelationships with the other DNA tumor viruses of the adeno and papilloma groups, but are often more productive and structurally smaller and simpler than the others and thus lend themselves to biochemical and biophysical research. Generally, one of the biochemist's most important aims is to produce the phenomenon he is studying under the simplest possible conditions. Thus there was another important advance when it became possible to replace tests on baby hamsters by the inoculation of cultured cells from young or embryonic hamsters (492, 302, 281). It is now possible to determine in this manner the frequency with which viral infection leads to cytocidal, as compared to transformational, response. It seems that under optimal conditions up to 5% of embryo hamster cells and 0.5% of rat embryo cells infected with about 1000 PFU of polyoma virus become transformed from producing virus to showing a changed appearance, altered surface properties, and transplantability (514). Specifically, transformed chick fibroblasts or baby hamster kidney cells, two cell types frequently used, lose their typical contact inhibition properties which keep normal cells apart and thus arrest cell division when a single cell layer has formed on a glass surface. Cells so transformed form easily identifiable clusters (Fig. 12-1).

As yet, no good, simple test for malignancy seems to exist, and the term "transformation" appears to be somewhat ambiguous in that sense. Transformed cells generally grow and divide indefinitely, but do not behave like malignant tumor cells when they are implanted in an animal. After long culture under varying conditions, however, transformed cells may develop the characteristics of malignancy, and then on implantation become able to produce tumors which invade neighboring tissues, including blood vessels, and thus tend to spread throughout the animal by metastasis (492, 302, 281).

Attempts to find infectious virus in transformed cells have been generally unsuccessful. Also, tests with antibodies to virus capsid proteins are usually negative, indicating that capsid proteins are not made. On the other hand, several virus-associated antigens can be detected in these cells. One of them, the T (tumor) antigen, is normally produced in untransformed infected cells before capsid proteins appear. Another antigen, the transplantation antigen, is not detected during cytocidal infection (161). A recent advance in the techniques of studying transformed cells is the recognition that the fusion of cells of different species (heterokaryons) caused by paramyxoviruses (e.g., Sendai virus; see Section VIII,J) leads to production of infective virus from transformed cells (497).

That viral DNA, for instance, that of SV40 (379), becomes integrated into the host's genome has been well established, and with adenovirus type 12 this seems to occur within the first 24 hr after infection of baby hamster kidney cells (83, 136). It seems that not the entire DNA, but possibly only half of it, needs to become integrated. The molecular mechanism of integration is not known, but it may well be similar to that of the integration of temperate phages into bacterial DNA (see Chapter XIII). It should be recalled that base composition and nearest neighbor analyses, as well as hybridization studies with the DNA's of papova viruses, have indicated considerable similarities among the various oncogenic viruses of these groups, as well as of the host's DNA as compared to that of the viruses (see Section VI,D) (221). It has also been recognized that there is appreciable sedimentation heterogeneity among the DNA molecules of polyoma and SV40 (within the circular supercoiled fraction), and the possibility that the oncogenic molecules may be defective is being investigated.

The similarities of sequences in the DNA of the host and of these classes of viruses probably account for the ease with which doubly transformed cells can be produced by the sequential infection with two viruses, such as polyoma and SV40, or adenovirus and SV40. Such doubly transformed cells show the presence of the T antigen of each virus. They may have greatly enhanced oncogenic activity.

Another interesting group of oncogenic agents are the mixed virions of various adenoviruses with SV40. They represent particles which contain DNA, and thus genomes, from the two types of parental virus that were applied to the cells (366, 155, 215).

Small amounts of infectious DNA can be isolated from SV40 transformed cells. No infective viral DNA has been found on adenovirus transformation, but the presence of appreciable amounts of virus-specific messenger RNA has been demonstrated; it seems to be transcribed from 0.001% of the cell's DNA (142).*

In conclusion: Papilloma virus preparations are always oncogenic although they frequently produce benign tumors; the adenoviruses and the polyoma and related groups can be cytocidal or harmless, they can also produce benign tumors, and the polyoma viruses can

* Two recent studies indicate that less of the viral genome becomes transcribed in transformed cells than during lytic infection (545, 547).

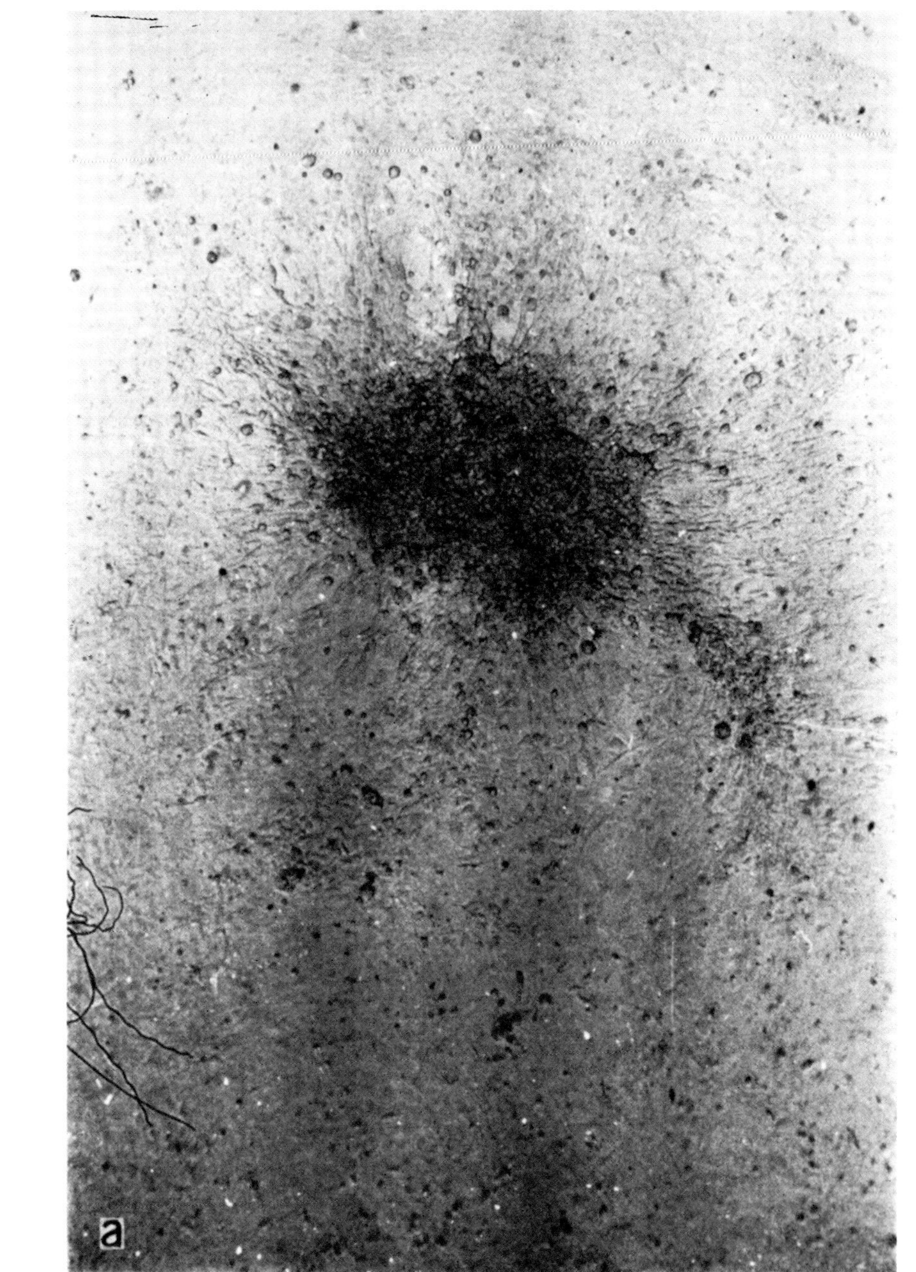
a

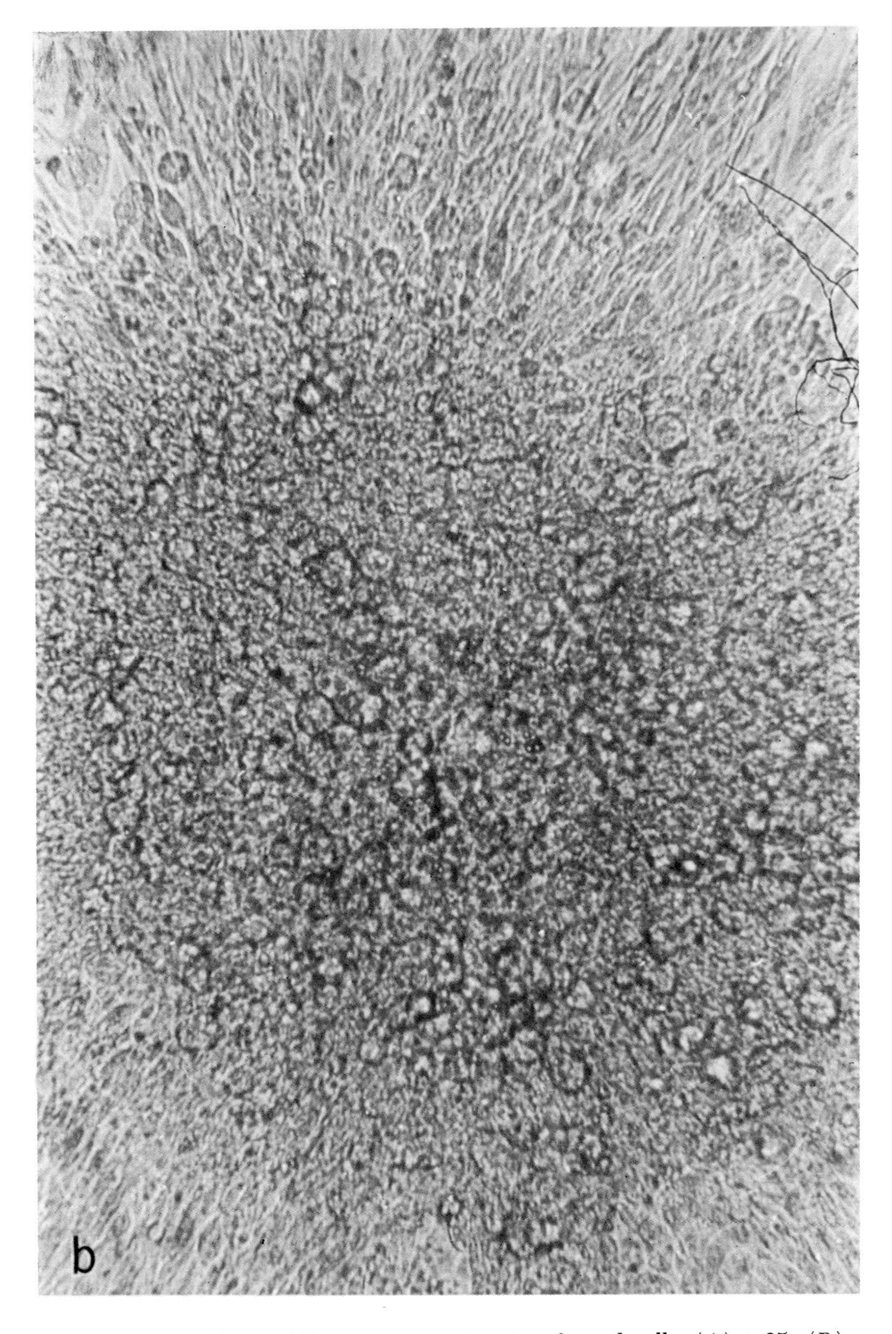

FIG. 12-1. Focus of Rous sarcoma virus transformed cells. (A) ×25; (B) ×100. (Courtesy of H. Rubin.)

produce malignant tumors in newborn rodents. For all of these viruses there is evidence of one kind or other that the viral DNA becomes in part integrated into the host's genome. This very active research field has recently been reviewed by Eckhard (97).

C. The RNA-Containing Tumor Viruses

The first clear identification of an oncogenic tumor virus was the finding by Rous in 1911 that a virus was the causative agent of a chicken sarcoma. Since then, but particularly since about 1950, many other viruses were detected which caused either solid tumors or tumors related to the lymphocyte system (leukemias) in birds and in mice. All these viruses proved to be related more or less closely in chemical and biological respects (Table 12.1). The Bittner

TABLE 12.1
The RNA Tumor Viruses (Leukoviruses)

Virus designation	*Disease and hosts*	*References*
Rous sarcoma (Bryan, Schmidt-Ruppin, etc.)	Solid tumors (birds)	364
Avian myeloblastosis	Bird leukemias, tumors of hemopoietic system	29
Avian leukosis	Bird leukemias, tumors of hemopoietic system	370, 372
Bittner mammary tumor	Carcinoma (mice)	36
Gross	Leukemias (mice)	156
Friend	Hemopoietic tumors (mice)	137
Graffi	Myeloid leukemia (mice)	154
Maloney	Lymphatic leukemia (mice)	287
Rauscher	Various leukemias (mice)	348
Mouse sarcoma	Sarcoma (mice)	172

mammary tumor virus (36) shows similar physical and biological properties and may therefore be included among these so-called leukoviruses, even though it does not seem to be serologically related to them.

Progress in the understanding of the Rous sarcoma virus came, as usual, after a test method for the virus became available, namely, the previously discussed formation and counting of foci of transformed cells on chick fibroblast monolayer culture (460). In contrast to the DNA tumor viruses, cells infected with Rous sarcoma virus (RSV) are able to produce virus while becoming transformed into tumor cells. More detailed studies have revealed a very complex situation, however. It appears that the commonly used Bryan high titer strain of RSV is one of several strains which fail to be infective in the standard chicken cells (164). This "pseudodefectiveness" is now regarded as a host range limitation rather than a viral-genetic phenomenon. Owing to this factor, the development of many RSV strains is markedly affected, in both positive and negative manner, by related latent viruses congenitally present in most chicken cells, such as RAV (Rous-associated virus) and RIF (resistance-inducing factor) (371). Apparently the simultaneous presence in a cell of both RSV and suitable levels of a latent avian leukosis virus such as RAV leads to infective RSV which shows the characteristic surface antigenic properties of RAV, the helper virus. It appears probable that some lipid-containing component of the viral envelope determines whether or not a certain particle is able to infect a certain cell type. In a recent very lucid review of this field, Vogt has defined the avian leukosis viruses as characterized by the same group-specific antigen and by the ability to act as helper for RSV (491). The avian leukosis viruses are transmitted through the egg. That they are not transmitted through the sperm, even though they occur in the testes, indicates that they are not integrated into the genome but remain cytoplasmic. These viruses cause leukemias in a relatively small but economically appreciable fraction of the chick population.

It appears that all cells infected by RSV become transformed and show the properties of malignancy (369). It has been suggested that this high oncogenicity, which is not attained by other viruses, is related to the defective aspects of RSV. However, this appears unlikely now, since strains of RSV which show no pseudodefective properties are also good transformers, and so is the Bryan high titer strain in susceptible host cells, such as quail cells.

That the tumors produced by RSV differ greatly in their virus content was found to be caused by either of two different mechanisms. On the one hand, tumors induced by large doses of virus in mature birds gradually lose most of their infective virus owing to an immunological response of the host, a reaction which can be suppressed by infecting the chicken embryos with avian leukosis, thus rendering them immunologically tolerant to the later infection with the related RSV (369, 370). On the other hand, infection at low multiplicities leads at times to noninfective tumors because the RSV may have failed to find sufficient helper virus to become coated (164, 469, 490, 365). By selecting conditions which quickly limit normal cell growth and multiplication it could be demonstrated that the characteristic changes of transformation occur within the first day after infection (372).

Conditions have recently been found to produce tumors, usually sarcomas, in mammals with certain strains of RSV (5). Again, newborn animals and particularly hamsters are susceptible. No infective RSV is produced by these tumors (162, 163) unless they are transferred back to chickens.

Certain observations in regard to RSV replication supply important hints concerning the mechanism of oncogenesis by an RNA virus. It appears that DNA synthesis must occur early during RSV infection, and that DNA function is needed throughout its replication (22), in distinct contrast to the replication of all nononcogenic RNA viruses. It has been proposed that an RNA-DNA hybrid might represent the equivalent of the replicative form (RF) of these tumor viruses (22). The formation of a double stranded DNA provirus, integrated into the host genome, has also been proposed (459). Hybridization experiments with host DNA and RSV RNA or another leukosis virus RNA have indicated that a small segment of these RNA's, particularly rich in A, is complementary to the host's DNA (165, 166).

CHAPTER XIII

THE BIOLOGY OF TEMPERATE PHAGES, LYSOGENY, AND TRANSDUCTION

The varying virus-host relationships have been compared earlier, with particular reference to the oncogenic response, in terms of historical events and the interaction of populations. For the nature of the temperate phages to be considered similarly, the role of the Negro in American history may represent an analogous situation. As will be discussed in detail, the temperate phages play an invisible role for many generations, replicating together with the cell genome without noticeable impact on the phenotype. However, temperate phages, induced by various agents or spontaneously, occasionally become frustrated by their invisibility or subjection and then turn, through a distinct biochemical decision, toward aggressive behavior, indistinguishable from that of virulent invading phages, and leading to paralysis of host functions and lysis and death of the host cell.

When temperate phages have turned intemperate and evoke the lytic response, their mode of action does not greatly differ from that of the virulent phages and needs discussion only insofar as it is necessary for the understanding of lysogeny. This discussion of the principal features of lysogeny and the related phenomena of general or specialized transduction, biological events of considerable interest, from both the biological-genetic and the molecular point of view, are brief and thus necessarily superficial. The reader is referred to other sources for a more thorough discussion of this complicated and challenging field (95, 177, 277).

Phage λ is probably the most intensively studied temperate coliphage. At least 18 genes have been identified in the λ genome by studying viral development in the lytic response, using a set of suppressor-sensitive conditional lethal nonsense mutants isolated by Campbell (53). Genes A to F were by such methods shown to be responsible for functions connected with phage tail protein production, and genes G to J are responsible for head production. R is the gene coding for the phage lysozyme (55). All of these thus are "late genes," in contrast to "early genes," such as N, which affects DNA replication. Considerable work has been done on the role of these various genes and the order in which they are transcribed. It has also become evident that different genes are transcribed from one or the other strand, and thus in the opposite direction (Fig. 13-1) (458).

The plaques produced by temperate phages usually show turbid centers which are due to the secondary growth of the lysogenized bacteria. Mutants which are unable to lysogenize are clearly obvious as clear plaques, and these c mutants (for clear) have played a decisive role in the elucidation of the mechanism of lysogeny (233).

It is now evident that lysogeny is the consequence of two main events: the repression of the typical viral functions, transcription, replication and translation on the one hand, and integration of the phage genome into that of the host on the other. Thus, in addition to the developmental genes discussed above, there exist the lysog-

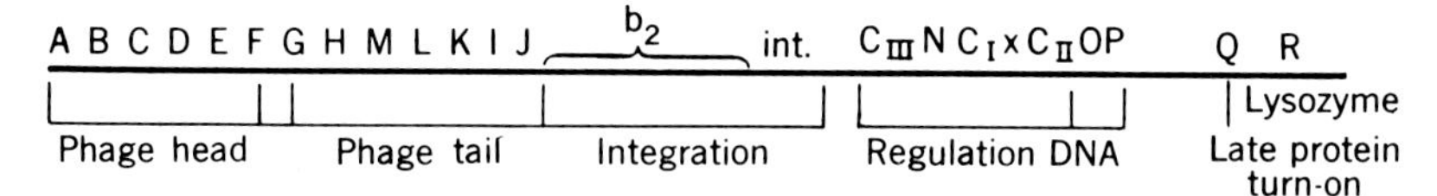

FIG. 13-1. Vegetative genetical map of λ phage. Genes A–R are required for the production of λ phage in lytic growth. Genes c_I, c_{II}, c_{III} are concerned with repression of lytic functions. The int gene and at least part of the b2 region are concerned with prophage integration (53, 95). The author wishes to thank Reinhold Book Corporation for permission to reproduce this diagram.

eny genes involved in repression, c_I, c_{II}, and c_{III}. Of these genes, c_I seems to be the only one essential for stable lysogeny. Thus it is particularly significant that the c_I repressor has recently been isolated and identified as a cytoplasmic protein. The functional identification of this gene product is its ability to bind to, and cosediment with, λ DNA. As expected, the c_I repressor does not bind to the DNA of λi 434, a mutant which is not homologous with λ in the region of c_I (345). The repression of gene expression is believed to be the consequence of this DNA-repressor interaction. It appears probable that the repressor prevents the transcription of one or more essential early genes probably related to DNA replication. The physicochemical mechanisms for specific binding of proteins on particular sites of nucleic acids which is encountered with these repressors, as well as with viral coat proteins (453) (see Section IX,B3), is not understood. The elucidation of these reactions represents one of the important challenges for future research.

It now seems well established that the second important stage of lysogenization represents true integration of the phage genome into the host DNA, and not only attachment. It appears that only circular supercoiled DNA can become integrated, and that this takes place through specific pairing of a homologous region of phage and host DNA (aa′, bb′). The incorporation of the phage genome occurs by reciprocal recombination in this region. Through circular permutation the gene order in the integrated linear genome is different from that in the virus, the late functions from both ends (R and A–F) becoming adjacent. This is obviously the result of the circular molecule opening at a site other than that represented by the cohesive ends of the phage DNA (Fig. 13-2, 6-9, 10) (54, 362).

Various pieces of genetic evidence clearly support the model in which λ DNA is inserted between the Gal and Bio markers of the host's DNA. It also seems clear that the phage provides the enzyme(s) necessary to achieve this integrative recombination. However, it is not yet clear whether the specificity of recombination is a consequence of base-pairing homology or whether it is contributed by the enzyme(s).

The integration of the phage genome at a specific locus obviously renders the bacterium unable to accommodate a related temperate

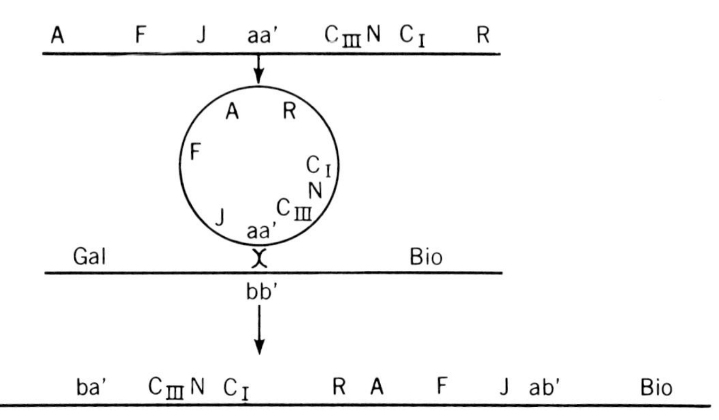

FIG. 13-2. The Campbell Model for Prophage Integration. The linear λ DNA molecule circularizes by joining cohesive sites, and a reciprocal recombination between the phage and host recognition regions (aa′ and bb′, respectively) provides for linear insertion of the viral DNA into that of the host. Recombinant recognition regions (ba′ and ab′) at either end of the prophage result (54, 362). The author wishes to thank Reinhold Book Corporation for permission to reproduce this diagram.

phage, and thus it produces "immunity." This is, as expected, a very specific type of immunity, since phages as closely related to one another as are λ, 434, 82, and 21, all of which have the same cohesive ends and are able to undergo genetic recombination, differ in *immunity specificity* (94). In other words, these phages become integrated at different sites of the bacterial genome.

Immunity is clearly different from the resistance which protects certain bacterial strains against infection by some virulent phages, a property of the cell wall rather than the genome.

The reversal of the process of lysogenization is the *induction* of vegetative phage development (278). This requires release from repression and excision of the phage-derived DNA. Of the various agents which can initiate this series of events, UV irradiation and treatment with mitomycin C are the most effective. The consequences of induction are the release, transcription, and replication of phage DNA, the production of viral proteins, the maturation of phage particles, and cell lysis.

Occasionally (e.g., one out of 10^6 particles) the induction leads to a *defective phage,* and it has become evident that this is due to an excision error leading to a deficit of original phage DNA compensated for by a piece of host DNA (17). Frequently compensation is not quantitative, and defective λ particles can thus show buoyant densities different from that of the typical phage. The specificity of the locus of integration is proved by the fact that such λ dg (defective galactose) particles always contain part of the adjacent host cistrons (galactose). Particles containing biotin markers of the host can also be detected (141), and other temperate phages carry parts of the neighboring genes of their respective attachment sites. Such defective viruses are generally infective only in combination with a related helper virus which serves to substitute the missing viral functions. The existence of these defective viruses has supplied a most useful tool in experimental virology (193).

General transduction and lysogeny appear to be related phenomena. Apparently a few temperate phages are not confined to a single highly specific attachment site, but can become integrated anywhere on the host genome. The resulting phenomenon of general transduction was first discovered with PLT22, which grows on *Salmonella typhimurium* (533). It was studied to best advantage with P1 because this phage grows on *E. coli*, the genetical map of which was known in some detail (267). The particular interest of these phages lies in their also tending to become faultily excised and then carry more or less of the host's genome, or even other prophages such as λ, into another host cell. P1 transducts different coli markers with roughly similar frequencies (e.g., 1 per 100,000 infective particles). The frequency of cotransduction of several markers is, as might be expected, a measure of their distance on the hosts genome (216). Thus transduction has proved to be a powerful tool in the mapping of bacterial genes. Surprisingly, as much as 2% of the *E. coli* genome can be transduced in one phage particle, even though this corresponds to two thirds of the normal DNA complement of that phage. Since transducing phages generally have the typical DNA complement, as indicated by their buoyant density, they must obviously be deficient in phage DNA (18). It is thus not surprising that the transducing particles do not cause infection, either lysogenic or lytic. This is a new natural phenomenon, subjugation and exploitation of the parasite by the host.

The bacteria have acquired a new means of achieving genetic recombination while emasculating the potentially harmful viral agent.

The phenomena described must definitely be regarded as marginal in terms of the domain of virology. There are no sharp lines distinguishing cellular and viral DNA. Parts of the genome of many bacteria are transferred under normal circumstances by either of two mechanisms: transformation or bacterial conjugation. Plasmids and episomes, nonchromosomal genes, are also transferred in this manner. They include the sex factors and the bacteriocinogenic factors, most of which have some properties in common with bacteriophages. The special role of the lysogenic and particularly the transducing phages, as far as bacterial genetics is concerned, is their ability to transfer almost any part of the bacterial chromosome. As far as the concept of viruses is concerned, it would appear that only the protein coat potential distinguishes viruses from host genome components. It thus appears advisable to retain this characteristic in all definitions of viruses, at least insofar as their genome is of DNA nature. Yet there is an obvious continuity of host, guest, and usurper, each dependent on the other.

The questions concerning the evolutionary origin(s) of viruses have not been resolved. Are viruses parasites derived from microorganisms which have lost many of their capabilities, or are they cellular organelles which have acquired the capability of extracellular existence? There are many indications that the oncogenic, lysogenic, and transducing viruses are very closely related to their hosts. It appears quite possible that the term virus covers particles with similar properties but with quite different origins.

BIBLIOGRAPHY

1. Aach, H. G. (1963). Electrophoretische Untersuchungen an Mutanten des Phagen øX174. *Z. Naturforsch.* **18b**, 290.
2. Aach, H. G., Funatsu, G., Nirenberg, M. W., and Fraenkel-Conrat, H. (1964). Further attempts to characterize products of TMV-RNA-directed protein synthesis. *Biochemistry* **3**, 1362.
3. Ackers, G. K., and Steere, R. L. (1967). Molecular sieve methods. *In* "Methods in Virology," Vol. II (K. Maramorosch and H. Koprowski, eds.), p. 325. Academic Press, New York.
4. Adams, J. M. (1968). On the release of the formyl group from nascent protein. *J. Mol. Biol.* 33, 571.
5. Ählstrom, C. G., and Forsby, N. (1962). Sarcomas in hamsters after infection with Rous chicken tumor material. *J. Exptl. Med.* **115**, 839.
6. Albertsson, P. (1967). Two phase separation of viruses. *In* "Methods in Virology," Vol. II (K. Maramorosch and H. Koprowski, eds.), p. 303. Academic Press, New York.
7. Anderer, F. A. (1959). Reversible Denaturierung des Proteins aus Tabakmosaikvirus. *Z. Naturforsch.* **14b**, 642.
8. Anderer, F. A. (1963). Recent studies on the structure of TMV. *Advan. Protein Chem.* **18**, 1.
9. Anderer, F. A., Uhlig, H., Weber, E., and Schramm, G. (1960). Primary structure of the protein of tobacco mosaic virus. *Nature* **186**, 922.
10. Anderer, F. A., Wittmann-Liebold, B., and Wittmann, H. G. (1965). Weitere Unterschungen zur Aminosäuresequenz des Proteins im Tabakmosaikvirus. *Z. Naturforsch.*, **20b**, 1203.
11. Anderer, F. A., and Schlumberger, H. D. (1966). Cross reactions of antisera against the terminal amino acid and dipeptide of tobacco mosaic virus. *Biochim. Biophys. Acta* **115**, 222.
12. Anderer, F. A., Schlumberger, H. D., Koch, M. A., Frank, H., and Eggers, H. J. (1967). Structure of Simian virus 40. II. Symmetry and components of the virus particle. *Virology* **32**, 511.
13. Anderer, F. A., Koch, M. A., Schlumberger, H. D., and Frank, H. (1968). Structure of Simian virus 40. III. Alkaline degradation of the virus particle. *Virology* **34**, 452.
14. Ansevin, A. T., Stevens, C. L., and Lauffer, M. A. (1964). Polymerization-depolymerization of tobacco mosaic virus proteins. III. Changes in ionization and in electrophoretic mobility. *Biochemistry* 3, 1512.
15. Aposhian, H. V., and Kornberg, A. (1962). Enzymatic synthesis of deoxyribonucleic acid. IV. The polymerase formed after T2 bacteriophage infection of *Escherichia coli,* a new enzyme. *J. Biol. Chem.* **237**, 519.
16. Appleyard, G., Hume, V. B. M., and Westwood, J. C. N. (1965). The effect of thiosemicarbazones on the growth of rabbitpox virus in tissue culture. *Ann. N.Y. Acad. Sci.* **130**, 92.

17. Arber, W. (1960). Transduction of chromosomal genes and episomes in *Escherichia coli*. *Virology* **11**, 273.
18. Arber, W., Kellenberger, G., and Weigle, J. (1957). La défectuosité du phage λ transducteur. *Schweiz. Z. Allgem. Pathol. Bakteriol.* **20**, 659.
19. Argetsinger, J. E., and Gussin, G. N. (1966). Intact RNA from defective particles of bacteriophage R17. *J. Mol. Biol.* **21**, 421.
20. Bachrach, H. L., and Woude, G. F. van de (1968). Amino acid composition and C-terminal sequence of foot and mouth disease virus protein. *Virology* **34**, 282.
21. Backus, R. C., and Williams, R. C. (1950). The use of spraying methods and of volatile suspending media in the preparation of specimens for electron microscopy. *J. Appl. Phys.* **20**, 11.
22. Bader, J. P. (1965). The requirement for DNA synthesis in the growth of Rous sarcoma and Rous-associated viruses. *Virology* **26**, 253.
23. Baltimore, D. (1968). Structure of the poliovirus replicative intermediate RNA. *J. Mol. Biol.* **32**, 359.
24. Bancroft, J. B., and Hiebert, E. (1967). Formation of an infectious nucleoprotein from protein and nucleic acid isolated from a small spherical virus. *Virology* **32**, 354.
25. Bancroft, J. B., Wagner, G. W., and Bracker, C. E. (1968). The self-assembly of a nucleic acid-free pseudo-top component for a small spherical virus. *Virology* **36**, 146.
25a. Bancroft, J. B., Hiebert, E., Rees, M. W., and Markham, R. (1968). Properties of cowpea chlorotic mottle virus, its protein and nucleic acid. *Virology* **34**, 224.
26. Bassel, B. A., Jr. (1968). Cell free protein synthesis directed by Qβ RNA and by two separate fragments. *Proc. Natl. Acad. Sci. U.S.* **60**, 321.
27. Bassel, B. A., Jr., and Spiegelman, S. (1967). Specific cleavage of Qβ RNA and identification of the fragment carrying the 3′-OH terminus. *Proc. Natl. Acad. Sci. U.S.* **58**, 1155.
28. Bawden, F. C., and Pirie, N. W. (1937). The isolation and some properties of liquid crystalline substances from solanaceous plants infected with three strains of tobacco mosaic virus. *Proc. Roy. Soc. (London)* **123**, 274.
29. Beard, J. W. (1963). Avian virus growth and their etiological agents. *Advan. Cancer Res.* **7**, 1.
30. Becker, A., Lyn, G., Getter, M., and Hurwitz, J. (1967). The enzymatic repair of DNA. II. Characterization of phage induced sealase. *Proc. Natl. Acad. Sci. U.S.* **58**, 1996.
31. Becker, Y., and Sarov, I. (1968). Electron miscroscopy of vaccinia virus DNA. *J. Mol. Biol.* **34**, 655.
32. Beijerinck, M. W. (1898). Ueber ein Contagium vivum fluidum als Ursache der Fleckenkrankheit der Tabaksblätter. *Zentr. Bakteriol. Parasitenk., Abt. II,* **5**, 27.

33. Bellamy, A. R., Shapiro, L., August, J. T., and Joklik, W. K. (1967). Studies on reovirus RNA. II. Characterization of reovirus genome RNA. *J. Mol. Biol.* **29**, 1.
34. Beukers, R., and Berends, W. (1960). Isolation and identification of the irradiation product of thymine. *Biochim. Biophys. Acta* **41**, 550.
35. Bishop, D. H. L., Claybrook, J. R., and Spiegelman, S. (1967). Electrophoretic separation of viral nucleic acids on polyacrylamide gels. *J. Mol. Biol.* **26**, 373.
35a. Bishop, D. H. L., Mills, D. R., and Spiegelman, S. (1968). The sequence of the 5′ terminus of a self-replicating variant of viral Qβ RNA. *Biochemistry* **7**, 3744.
35b. Bishop, D. H. L., Pace, N. R., and Spiegelman, S. (1967). The mechanism of replication: A novel polarity reversal in the *in vitro* synthesis of Qβ RNA and its complement. *Proc. Natl. Acad. Sci. U.S.* **58**, 1790.
36. Bittner, J. J. (1936). Some possible effects of nursing on the mammary gland tumor incidence in mice. *Science* **84**, 162.
37. Bockstahler, L. E., and Kaesberg, P. (1965). Isolation and properties of RNA from bromegrass mosaic virus. *J. Mol. Biol.* **13**, 127.
38. Bode, V. C., and Kaiser, A. D. (1965). Changes in the structure and activity of λ DNA in a superinfected immune bacterium. *J. Mol. Biol.* **14**, 399.
39. Bosch, L., Bonnet-Smits, E. M., and van Duin, J. (1967). *In situ* breakage of turnip yellow mosaic virus RNA and *in situ* aggregation of fragments. *Virology* **31**, 453.
39a. Bové, J. M., Bové, C., and Mocquot, B. (1968). Turnip yellow mosaic virus RNA synthesis *in vitro*: Evidence for native double-stranded RNA. *Biochim. Biophys. Res. Commun.* **32**, 480.
40. Boy de la Tour, E., and Kellenberger, E. (1965). Aberrant forms of T-even phage head. *Virology* **27**, 222.
41. Brakke, M. K. (1960). Density gradient centrifugation and its application to plant viruses. *Advan. Virus Res.* **7**, 193.
42. Brakke, M. K. (1967). Density gradient centrifugation. *In* "Methods in Virology," Vol. II (K. Maramorosch and H. Koprowski, eds.), p. 93. Academic Press, New York.
43. Brammer, K. W. (1963). Chemical modification of viral ribonucleic acid. II. Bromination and iodination. *Biochim. Biophys. Acta* **72**, 217.
44. Braunitzer, G., Hobom, G., and Hannig, K. H. S. (1964). Ein einfaches Verfahren zur Selektionierung von Mutanten. *Z. Physiol. Chem.* **338**, 278.
45. Braunitzer, G., Asbeck, F., Beyreuther, K., Köhler, H., and von Wettstein, G. (1967). Die Konstitution des Hüllproteins des Bakteriophagen fd. *Z. Physiol. Chem.* **348**, 725.
46. Brenner, S., Streisinger, G., Horne, R. W., Champe, S. P., Barnett, L., Benzer, S., and Rees, M. W. (1958). Structural components of bacteriophage. *J. Mol. Biol.* **1**, 281.

47. Brenner, S., and Horne, R. W. (1959). A negative staining method for high resolution electron microscopy of viruses. *Biochim. Biophys. Acta* **34**, 103.
48. Brody, E., Coleman, L., Mackal, R. P., Werninghaus, B., and Evans, E. A., Jr. (1964). Properties of infectious DNA from T1 and λ bacteriophage. *J. Biol. Chem.* **239**, 285.
49. Brown, D. M., and Hewlins, M. J. E. (1968). The reaction between hydroxylamine and cytosine derivatives. *J. Chem. Soc.* (C), p. 1922.
50. Brown, D. M., Hewlins, M. J. E., and Schell, P. (1968). The tautomeric state of N(4) hydroxy and of N(4) amino cytosine derivatives. *J. Chem. Soc.* (C), p. 1925.
51. Bruening, G., and Agrawal, H. O. (1967). Infectivity of a mixture of cowpea mosaic virus ribonucleoprotein components. *Virology* **31**, 217.
52. Butler, P. J. G., Harris, J. I., Hartley, B. S., and Leberman, R. (1967). Reversible blocking of peptide amino groups by maleic anhydride. *Biochem. J.* **103**, 78P.
53. Campbell, A. (1961). Sensitive mutants of bacteriophage λ. *Virology* **14**, 22.
54. Campbell, A. (1962). Episomes. *Advan. Genet.* **11**, 101.
55. Campbell, A., and del Campillo-Campbell, A. (1963). Mutant of λ bacteriophage producing a thermolabile endolysin. *J. Bacteriol.* **85**, 1202.
56. Capecchi, M. R. (1966). Cell-free protein synthesis programmed with R17 RNA; identification of two phage proteins. *J. Mol. Biol.* **21**, 733.
57. Caro, L. G., and Schnös, M. (1966). The attachment of the male-specific bacteriophage f1 to sensitive strains of *Escherichia coli*. *Proc. Natl. Acad. Sci. U.S.* **56**, 126.
58. Caspar, D. L. D. (1956). The radial density distribution in the tobacco mosaic virus particle. *Nature* **177**, 928.
59. Caspar, D. L. D. (1963). Assembly and stability of the TMV particle. *Advan. Protein Chem.* **18**, 37.
60. Caspar, D. L. D., and Klug, A. (1962). Physical principles in the construction of regular viruses. *Cold Spring Harbor Symp. Quant. Biol.* **27**, 1.
61. Cech, M. (1967). On the role of chloroplasts in TMV reproduction. *Phytopathol. Z.* **59**, 72.
62. Cerda-Olmedo, E., Hanawalt, P. C., and Guerola, N. (1968). Mutagenesis of the replication point by nitrosoguanidine. *J. Mol. Biol.* **33**, 705.
63. Chandra, P., Wacker, A., Süssmuth, R., and Lingens, F. (1967). Wirkung von 1-nitroso-3 nitro-1 methyl-guanidin auf die Matrizenaktivität der Polynucleotide bei der zellfreien Protein Synthese. *Z. Naturforsch.* **22b**, 512.
64. Choppin, P. W., and Philipson, L. (1961). The inactivation of enterovirus activity by the sulfhydryl reagent, *p*-chloromercuribenzoate. *J. Exptl. Med.* **113**, 713.

65. Clark, J. M., Chang, A. Y., Spiegelman, S., and Reichmann, M. E. (1965). The *in vitro* translation of a monocistronic message. *Proc. Natl. Acad. Sci. U.S.* **54**, 1193.
66. Cohen, L. A. (1968). Group-specific reagents in protein chemistry. *Ann. Rev. Biochem.* **37**, 683.
67. Colby, C., and Chamberlin, M. J. (1969). The specificity of interferon induction in chick embryo cells by helical RNA. *Proc. Natl. Acad. Sci. U.S.* (in press).
68. Cooper, P. D. (1968). A genetic map of poliovirus temperature sensitive mutants. *Virology* **35**, 584.
69. Crawford, L. (1968). *In* "Molecular Basis of Virology" (H. Fraenkel-Conrat, ed.), p. 393. Reinhold Publishing Corp., New York.
70. Crick, F. H. C. (1962). The genetic code. *Sci. Am.* **207**, 66.
71. Crick, F. H. C., Barnett, L., Brenner, S, and Watts-Tobin, R. J. (1961). General nature of the genetic code for proteins. *Nature* **192**, 1227.
72. Crocker, T., Pfendt, E., and Spendlove, R. (1964). Poliovirus: Growth in non-nucleate cytoplasm. *Science* **145**, 401.
73. Crowell, R. L. (1966). Specific cell-surface alteration by enteroviruses as reflected by viral attachment interference. *J. Bacteriol.* **91**, 198.
74. Cummings, D. J. (1963). Subunit basis of head configurational changes in T2 bacteriophage. *Biochim. Biophys. Acta* **68**, 472.
74a. Dahlberg, J. E. (1968). Terminal sequences of bacteriophage RNAs. *Nature* **220**, 548.
75. Dales, S. (1965). Penetration of animal viruses into cells. *Progr. Med. Virol.* **7**, 1.
76. Dales, S., and Choppin, P. W. (1962). Attachment and penetration of influenza virus. *Virology* **18**, 489.
77. Dales, S., Gomatos, P. J., and Hsu, K. C. (1965). The uptake and development of reovirus in strain L cells followed with labeled viral ribonucleic acid and ferritin-antibody conjugates. *Virology* **25**, 193.
78. Davison, P. F., Freifelder, D., Hede, R., and Levinthal, C. (1961). The structural unity of the DNA of T2 bacteriophage. *Proc. Natl. Acad. Sci. U.S.* **47**, 1123.
79. Delbrück, M. (1940). The growth of bacteriophage and lysis of the host. *J. Gen. Physiol.* **23**, 643.
80. Delbrück, M., and Bailey, W. T., Jr. (1946). Induced mutations in bacterial viruses. *Cold Spring Harbor Symp. Quant. Biol.* **11**, 33.
81. Diener, T. O., and Raymer, W. B. (1967). Potato spindle tuber virus: A plant virus with properties of a free nucleic acid. *Science* **158**, 378.
82. Dingman, C. W., and Peacock, A. C. (1968). Analytical studies on nuclear ribonucleic acid using polyacrylamide gel electrophoresis. *Biochemistry* **7**, 659.
83. Doerfler, W. (1968). The fate of the DNA of adenovirus type 12 in baby hamster kidney cells. *Proc. Natl. Acad. Sci. U.S.* **60**, 636.
84. Doerfler, W., and Hogness, D. S. (1968). The strands of DNA from

λ and related bacteriophages: Isolation and characterization. *J. Mol. Biol.* **33**, 635.

85. Dorne, B., and Hirth, L. (1968). Influence d'organo-mercuriels sur la configuration du virus du rabougrissement buissonneux de la tomate. *Compt. Rend.* **267**, 127.
86. Drzeniek, R., Seto, F. T., and Rott, R. (1966). Characterization of neuraminidases from myxoviruses. *Biochim. Biophys. Acta* **128**, 547.
87. Duesberg, P. H. (1968). The RNA's of influenza virus. *Proc. Natl. Acad. Sci. U.S.* **59**, 930.
88. Duesberg, P. H. (1968). Physical properties of Rous sarcoma virus RNA. *Proc. Natl. Acad. Sci. U.S.* **60**, 1511.
89. Duesberg, P. H., Robinson, H. L., Robinson, W. S., Huebner, R. J., and Turner, H. C. (1968). Proteins of Rous sarcoma virus. *Virology* **36**, 73.
90. Dulbecco, R., and Vogt, M. (1953). Some problems of animal virology as studied by the plaque technique. *Cold Spring Harbor Symp. Quant. Biol.* **18**, 273.
91. Dunn, D. B., and Smith, J. D. (1958). The occurrence of 6-methylaminopurine in deoxyribonucleic acids. *Biochem. J.* **68**, 627.
92. Dunnebacke, T. H., and Kleinschmidt, A. K. (1967). Ribonucleic acid from reovirus as seen in protein monolayers by electron microscopy. *Z. Naturforsch.* **22b**, 159.
93. Dyson, R. D., and van Holde, K. E. (1967). An investigation of bacteriophage λ, its protein ghosts and subunits. *Virology* **33**, 559.
94. Echols, H., and Gingery, R. (1968). Properties of λ mutants defective in vegetative recombination and integration. *J. Mol. Biol.* **34**, 239.
95. Echols, H., and Joyner, A. (1968). The temperate bacteriophage. *In* "Molecular Basis of Virology" (H. Fraenkel-Conrat, ed.), p. 526. Reinhold Publishing Corp., New York.
96. Eckert, E. A. (1966). Envelope protein(s) derived from influenza virus. *J. Bacteriol.* **91**, 1907.
97. Eckhart, W. (1968). Transformation of animal cells by oncogenic DNA viruses. *Physiol. Rev.* **48**, 513.
98. Edgar, R. S. (1966). *In* "Phage and the Origins of Molecular Biology" (Cairns *et al.*, eds.), p. 166. Cold Spring Harbor Laboratory of Quantitative Biology. Cold Spring Harbor, New York.
99. Edgar, R. S., and Wood, W. B. (1966). Morphogenesis of bacteriophage T4 in extracts of mutant-infected cells. *Proc. Natl. Acad. Sci. U.S.* **55**, 498.
100. Edgar, R. S., and Lielausis, I. (1968). Some steps in the assembly of bacteriophage T4. *J. Mol. Biol.* **32**, 263.
101. Elford, W. J. (1938). The sizes of virus and bacteriophages and methods for their determination. *In* "Handbuch der Virusforschung" R. Doerr and C. Hallauer, eds.). Wien, Springer.
102. Ellis, E. L., and Delbrück, M. (1939). The growth of bacteriophage. *J. Gen. Physiol.* **22**, 365.

103. Farid, S. A. A., and Kozloff, L. M. (1968). Number of polypeptide components in bacteriophage T2L contractible sheath. *J. Virol.* **2**, 308.
104. Fazekas de St. Groth, S. (1948). Viropexis, the mechanism of influenza virus infection. *Nature* **162**, 294.
105. Fazekas de St. Groth, S., and Webster, R. G. (1963). The neutralization of animal viruses. IV. Parameters of the influenza virus-antibody system. *J. Immunol.* **90**, 151.
106. Feix, G., Slor, H., and Weissmann, C. (1967). Replication of viral RNA, XIII. The early product of phage RNA synthesis *in vitro*. *Proc. Natl. Acad. Sci. U.S.* **57**, 1401.
106a. Feix, G., Pollet, R., and Weissmann, C. (1968). Replication of viral RNA. XVI. Enzymatic synthesis of infectious viral RNA with non-infectious Qβ minus strands as template. *Proc. Natl. Acad. Sci. U.S.* **59**, 145.
107. Feldman, M. Ya. (1965). Formation of methylene bridges in the reaction of ribonucleic acid with formaldehyde. *Biokhimiya* **30**, 203.
108. Feldman, M. Ya. (1967). Reaction of formaldehyde with nucleotides and RNA. *Biochim. Biophys. Acta* **149**, 20.
109. Fenner, F. (1968). "The Biology of Animal Viruses," Vol I. Academic Press, New York.
110. Fiers, W., and Sinsheimer, R. L. (1962). The structure of the DNA of bacteriophage φX174. I. The action of exopolynucleotidases. *J. Mol. Biol.* **5**, 408.
111. Finch, J. T. (1965). Preliminary X-ray diffraction studies on tobacco rattle and barley stripe mosaic viruses. *J. Mol. Biol.* **12**, 612.
112. Finch, J. T., and Holmes, K. C. (1967). Structural studies of viruses. *In* "Methods in Virology," Vol. III (K. Maramorosch and H. Koprowski, eds.), p. 352. Academic Press, New York.
112a. Finch, J. T., and Bancroft, J. B. (1968). Structure of the reaggregated protein shells of two spherical viruses. *Nature* **220**, 815.
113. Fine, R., and Murakami, W. T. (1968). Protein composition of polyoma virus. *J. Mol. Biol.* **36**, 167.
114. Fraenkel-Conrat, H. (1954). Reaction of nucleic acid with formaldehyde. *Biochim. Biophys. Acta* **15**, 308.
115. Fraenkel-Conrat, H. (1956). The role of the nucleic acid in the reconstitution of active tobacco mosaic virus. *J. Am. Chem. Soc.* **78**, 882.
116. Fraenkel-Conrat, H. (1957). Degradation and structure of tobacco mosaic virus. *Federation Proc.* **16**, 810.
117. Fraenkel-Conrat, H. (1957). Degradation of tobacco mosaic virus with acetic acid. *Virology* **4**, 1.
118. Fraenkel-Conrat, H. (1957). Methods for investigating the essential groups for enzyme activity. *In* "Methods of Enzymology," Vol. IV (S. Colowick and N. O. Kaplan, eds.), p. 247. Academic Press, New York.
119. Fraenkel-Conrat, H. (1959). The masked —SH group in tobacco mo-

saic virus protein. *In* "Symposium on Sulfur in Proteins" (R. Benesch, ed.), p. 339. Academic Press, New York.
120. Fraenkel-Conrat, H. (1961). Chemical modification of viral ribonucleic acid. I. Alkylating agents. *Biochim. Biophys. Acta* **49**, 169.
121. Fraenkel-Conrat, H. (1965). Structure and function of virus proteins and of viral nucleic acid. *In* "The Proteins," Vol. III (H. Neurath, ed.), p. 99. Academic Press, New York.
122. Fraenkel-Conrat, H. (1968). *In* "Molecular Basis of Virology" (H. Fraenkel-Conrat, ed.), p. 134. Reinhold Publishing Corp., New York.
123. Fraenkel-Conrat, H., and Colloms, M. (1967). Reactivity of tobacco mosaic virus and its protein towards acetic anhydride. *Biochemistry* **6**, 2740.
124. Fraenkel-Conrat, H., and Rueckert, R. R. (1967). Analysis of protein constituents of viruses. *In* "Methods in Virology," Vol. III (K. Maramorosch and H. Koprowski, eds.), p. 1. Academic Press, New York.
125. Fraenkel-Conrat, H., and Singer, B. (1957). Virus reconstitution. II. Combination of protein and nucleic acid from different strains. *Biochim. Biophys. Acta* **24**, 540.
126. Fraenkel-Conrat, H., and Singer, B. (1959). Reconstitution of tobacco mosaic virus. III. Improved methods and the use of mixed nucleic acids. *Biochim. Biophys. Acta* **33**, 359.
127. Fraenkel-Conrat, H., and Singer, B. (1964). Reconstitution of tobacco mosaic virus. IV. Inhibition by enzymes and other proteins, and use of polynucleotides. *Virology* **23**, 354.
128. Fraenkel-Conrat, H., and Williams, R. C. (1955). Reconstitution of active tobacco mosaic virus from the inactive protein and nucleic acid components. *Proc. Natl. Acad. Sci. U.S.* **41**, 690.
129. Fraenkel-Conrat, H., Singer, B., and Tsugita, A. (1961). Purification of viral RNA by means of bentonite. *Virology* **14**, 54.
130. Fraenkel-Conrat, H., Singer, B., and Veldee, S. (1958). The mechanism of plant virus infection. *Biochim. Biophys. Acta* **29**, 639.
130a. Fraenkel-Conrat, H., Singer, B., and Williams, R. C. (1957). Infectivity of viral nucleic acid. *Biochim. Biophys. Acta* **25**, 87.
131. Frankel, F. R. (1966a). The absence of mature phage DNA molecules from the duplicating pool of T-even infected *E. coli*. *J. Mol. Biol.* **18**, 109.
132. Frankel, F. R. (1966b). Studies on the nature of replicating DNA in T4-infected *Escherichia coli*. *J. Mol. Biol.* **18**, 127.
133. Franklin, R. E. (1956). X-ray diffraction studies of cucumber virus 4 and three strains of tobacco mosaic virus. *Biochim. Biophys. Acta* **19**, 203.
134. Franklin, R. E., Klug, A., and Holmes, K. C. (1957). X-ray diffraction studies of the structure and morphology of tobacco mosaic virus, 5. *In* "CIBA Foundation Symposium on the Nature of Viruses" (G. E. W. Wolstenholme and E. C. P. Millar, eds.). Little, Brown and Co., Boston.

135. Franze de Fernandez, M. T., Eoyang, L., and August, J. T. (1968). Factor fraction required for the synthesis of bacteriophage Qβ-RNA. *Nature* **219**, 588 (see also Nature **219**, 675).

136. Freeman, A. E., Black, P. H., Wolford, R., and Huebner, R. J. (1967). Adenovirus type 12—rat embryo transformation system. *J. Virol.* **1**, 362.

137. Friend, C. (1957). Cell free transmission in adult Swiss mice of a disease having the character of leukemia. *J. Exptl. Med.* **105**, 307.

138. Friesen, B. S., and Sinsheimer, R. L. (1959). Partition cell analysis of infective tobacco mosaic virus nucleic acid. *J. Mol. Biol.* **1**, 321.

139. Frist, R. H., Bendet, I. J., Smith, K. M., and Lauffer, M. A. (1965). The protein subunit of cucumber virus 4; degradation of viruses by succinylation. *Virology* **26**, 558.

140. Frost, R. R., Harrison, B. D., and Woods, R. D. (1967). Apparent symbiotic interaction between particles of tobacco rattle virus. *J. Gen. Virol.* **1**, 57.

141. Fuerst, C. R. (1966). Defective biotin-transducing mutants of bacteriophage λ. *Virology* **30**, 581.

142. Fujinaga, K., and Green, M. (1966). The mechanisms of viral carcinogenesis by DNA mammalian viruses: Viral-specific RNA in polyribosomes of adenovirus tumor and transformed cells. *Proc. Natl. Acad. Sci. U.S.* **55**, 1567.

143. Funatsu, G., and Fraenkel-Conrat, H. (1964). Location of amino acid exchanges in chemically evoked mutants of tobacco mosaic virus. *Biochemistry* **3**, 1356.

144. Funatsu, G., and Funatsu, M. (1968). Chemical studies on proteins from two tobacco mosaic virus strains. *In* Biochemical Regulation in Diseased Plants. *Phytopathol. Soc. Japan,* p. 1.

145. Funatsu, G., Tsugita, A., and Fraenkel-Conrat, H. (1964). Studies on the amino acid sequence of tobacco mosaic virus protein. V. Amino acid sequences of two peptides from tryptic digests and location of amide group. *Arch. Biochem. Biophys.* **105**, 25.

146. Gibbs, A. J., Nixon, H. L., and Woods, R. D. (1963). Properties of purified preparations of lucern mosaic virus. *Virology* **19**, 441.

147. Gierer, A., and Schramm, G. (1956). Die Infektiosität der Ribonukleinsäure des Tabakmosaikvirus. *Z. Naturforsch.* **11b**, 138 (also *Nature* **177**, 702).

148. Gierer, A., and Mundry, K. W. (1958). Production of mutants of TMV by chemical alteration of its ribonucleic acid *in vitro*. *Nature* **182**, 1457.

149. Glitz, D. G. (1968). The nucleotide sequence at the 3′-linked end of bacteriophage MS2 RNA. *Biochemistry* **7**, 927.

150. Glitz, D. G., Bradley, A., and Fraenkel-Conrat, H. (1968). Nucleotide sequences at the 5′-linked ends of viral RNAs. *Biochim. Biophys. Acta* **161**, 1.

151. Gordon, M. P., and Staehelin, M. (1959). Studies on the incorporation of 5-fluorouracil into virus nucleic acid. *Biochim. Biophys. Acta* **36**, 351.
152. Gottschalk, A. (1959). Chemistry of virus receptors. *In* "The Viruses" Vol. 3 (F. M. Burnet and W. M. Stanley, eds.), p. 51. Academic Press, New York.
153. Goulian, M., Kornberg, A., and Sinsheimer, R. L. (1967). Enzymatic synthesis of DNA, XXIV. Synthesis of infectious phage ϕX174 DNA. *Proc. Natl. Acad. Sci. U.S.* **58**, 2321.
154. Graffi, A., Bielkar, A., and Fey, F. (1956). Leukämieerzeugung durch ein filtrierbares Agens aus malignen Tumoren. *Acta Haematol.* **15**, 145.
155. Green, M., Piña, M., and Fujinaga, K. (1966). Biosynthetic modifications induced by DNA animal viruses. *Ann. Rev. Microbiol.* **20**, 189.
156. Gross, L. (1951). Spontaneous leukemia developing in C3H mice following inoculation in infancy with AK leukemia extracts of AK embryos. *Proc. Soc. Exptl. Biol. Med.* **76**, 27.
157. Grunberg-Manago, M., and Michelson, A. M. (1964). Polynucleotide analogues: Polyfluorouridylic acid and copolymers containing fluorouridylic acid. *Biochim. Biophys. Acta* **87**, 593.
158. Guha, A., and Szybalski, W. (1968). Fractionation of the complementary strands of coliphage T4 DNA based on the asymmetric distribution of the poly U and poly UG binding sites. *Virology* **34**, 608.
159. Gussin, G. N. (1966). Three complementation groups in bacteriophage R17. *J. Mol. Biol.* **21**, 435.
160. Guthrie, G. C., and Buchanan, J. (1966). Control of phage-induced enzymes in bacteria. *Federation Proc.* **25**, 864.
161. Habel, K. (1965). Specific complement-fixing antigens in polyoma tumors and transformed cells. *Virology* **25**, 55.
162. Hanafusa, H., and Hanafusa, T. (1966a). Determining factor in the capacity of Rous sarcoma virus to induce tumors in mammals. *Proc. Natl. Acad. Sci. U.S.* **55**, 532.
163. Hanafusa, H., and Hanafusa, T. (1966b). Analysis of the defectiveness of Rous sarcoma virus. IV. Kinetics of RSV production. *Virology* **28**, 369.
164. Hanafusa, H., Hanafusa, T., and Rubin, H. (1963). The defectiveness of Rous sarcoma virus. *Proc. Natl. Acad. Sci. U.S.* **49**, 572.
165. Harel, J., Harel, L., Golde, A., and Vigier, P. (1966a). Homologie entre genome du virus du sarcome de Rous (RSV) et genome cellulaire. *Compt. Rend.* **263**, 745.
166. Harel, L., Harel, J., Lacour, F., and Huppert, J. (1966b). Homologie entre genome du virus de la myeloblastose aviaire (AMV) et genome cellulaire. *Compt. Rend.* **263**, 616.
167. Harris, J. I., and Knight, C. A. (1952). Action of carboxypeptidase on TMV. *Nature* **170**, 613.
168. Harris, J. I., and Hindley, J. (1965). The protein subunit of turnip yellow mosaic virus. *J. Mol. Biol.* **13**, 894.

169. Hart, R. G. (1958). The nucleic acid fiber of the tobacco mosaic virus particle. *Biochim. Biophys. Acta* **28**, 457.
170. Haruna, I., and Spiegelman, S. (1965). Autocatalytic synthesis of a viral RNA *in vitro*. *Science* **150**, 885.
171. Haruna, I., and Spiegelman, S. (1965). Recognition of size and sequence by an RNA replicase. *Proc. Natl. Acad. Sci. U.S.* **54**, 1189.
172. Harvey, J. J. (1964). An unidentified virus which causes the rapid production of tumors in mice. *Nature* **204**, 1104.
173. Haschemeyer, R., Singer, B., and Fraenkel-Conrat, H. (1959). Two configurations of tobacco mosaic virus ribonucleic acid. *Proc. Natl. Acad. Sci. U.S.* **45**, 313.
174. Haselkorn, R., and Doty, P. (1961). The reaction of formaldehyde with polynucleotides. *J. Biol. Chem.* **236**, 2738.
175. Hattman, S., and Fukusawa, T. (1963). Host-induced modification of T-even phages due to defective glucosylation of their DNA. *Proc. Natl. Acad. Sci. U.S.* **50**, 297.
176. Haukenes, G., Harboe, A., and Mortensson-Egnund, K. (1965). A uronic and sialic acid free chick allantoic mucopolysaccharide sulfate which combines with influenza virus HI antibody to host material. *Acta Pathol. Microbiol. Scand.* **64**, 534.
177. Hayes, W. (1964). "The Genetics of Bacteria and Their Viruses." John Wiley and Sons, New York.
178. Heisenberg, M. (1966). Formation of defective bacteriophage particles by fr amber mutants. *J. Mol. Biol.* **17**, 136.
179. d'Hérelle, F. (1917). Sur un microbe invisible antagoniste des bacilles dysentériques. *Compt. Rend.* **165**, 373.
180. Herrmann, R., Schubert, D., and Rudolph, U. (1968). Self-assembly of protein subunits from bacteriophage fr. *Biochem. Biophys. Res. Commun.* **30**, 576.
181. Hershey, A. D. (1946). Mutation of bacteriophage with respect to type of plaque. *Genetics* **31**, 620.
182. Hershey, A. D., and Rotman, R. (1949). Genetic recombination between host-range and plaque-type mutants of bacteriophage in single bacterial cells. *Genetics* **34**, 44.
183. Hershey, A. D., and Chase, M. (1952). Independent functions of viral protein and nucleic acid in growth of bacteriophage. *J. Gen. Physiol.* **36**, 39.
184. Hershey, A. D., Burgi, E., and Ingraham, L. (1963). Cohesion of DNA molecules isolated from phage λ. *Proc. Natl. Acad. Sci. U.S.* **49**, 784.
185. Hershey, A. D., Burgi, E., and Davern, C. J. (1965). Preparative density-gradient centrifugation of molecular halves of λ DNA. *Biochem. Biophys. Res. Commun.* **18**, 675.
186. Hiebert, E., Bancroft, J. B., and Bracker, C. E. (1968). The assembly *in vitro* of some small spherical viruses, hybrid viruses, and other nucleoproteins. *Virology* **34**, 492.

186a. Hilleman, M. P. (1968). Interferon induction and utilization. *J. Cellular Physiol.* **71**, 43.

187. Hitchborn, H. J., and Hills, G. J. (1968). A study of tubes produced in plants infected with a strain of TYMV. *Virology* **35**, 50.

188. Hoffmann-Berling, H., Dürwald, H., and Beulke, I. (1963). Ein fädiger DNS-Phage (fd) und ein sphärischer RNS-Phage (fr) wirtsspezifisch für männliche Stämme von *E. coli.* III. Biologisches Verhalten von fd und fr. *Z. Naturforsch.* **18b**, 893.

189. Hofschneider, P. H., and Hausen, P. (1968). The replication cycle. *In* "Molecular Basis of Virology" (H. Fraenkel-Conrat, ed.), p. 169. Reinhold Publishing Corp., New York.

190. Hoggan, M. D., Blacklow, N. R., and Rowe, W. P. (1968). Studies of small DNA viruses found in various adenovirus preparations. *Proc. Natl. Acad. Sci. U.S.* **55**, 1467.

191. Hoggan, M. D., Shatkin, A. J., Blacklow, N. R., Koczot, F., and Rose, R. A. (1968). Helper-dependent infectious DNA from adeno virus associated virus. *J. Virol.* **2**, 850.

192. Hogness, D. S., and Simmons, J. R. (1964). Breakage of λdg DNA: Chemical and genetic characterization of each isolated half molecule. *J. Mol. Biol.* **9**, 411.

193. Hogness, D. S., Doerfler, W., Egan, J. B., and Black, L. W. (1966). The position and orientation of genes in λ and λdg DNA. *Cold Spring Harbor Symp. Quant. Biol.* **31**, 129.

194. Hohn, T. (1967). Self assembly of defective particles of the bacteriophage fr. *European J. Biochem.* **2**, 152.

195. Holland, J. J. (1961). Receptor affinities as major determinants of enterovirus tissue tropisms in humans. *Virology* **15**, 312.

196. Holland, J. J., McLaren, L. C., and Syverton, J. T. (1959). The mammalian cell-virus relationship. IV. Infection of naturally insusceptible cells with enterovirus ribonucleic acids. *J. Exptl. Med.* **110**, 65.

197. Holmes, F. O. (1929). Local lesions in tobacco mosaic. *Botan. Gaz.* **87**, 39.

198. Holoubek, V. (1962). Mixed reconstitution between protein from common tobacco mosaic virus and ribonucleic acid from other strains. *Virology* **18**, 401.

199. Holowczak, J. A., and Joklik, W. K. (1967). Studies on the proteins of vaccinia virus. 1. Structural proteins of virions and cores. 2. Kinetics of the synthesis of individual groups of structural proteins. *Virology* **33**, 717, 726.

200. Horiuchi, K., Lodish, H. F., and Zinder, N. D. (1966). Mutants of bacteriophage f2. *Virology* **28**, 438.

201. Horne, R. W. (1967). Electron microscopy of isolated virus particles and their components. *In* "Methods in Virology," Vol. III (K. Maramorosch and H. Koprowski, eds.), p. 522. Academic Press, New York.

202. Horne, R. W., and Waterson, A. P. (1960). A helical structure in mumps, Newcastle disease and Sendai viruses. *J. Mol. Biol.* **2**, 75.

203. Horne, R. W., Waterson, A. P., Wildy, P., and Farnham, A. E. (1960). The structure and composition of the myxoviruses. *Virology* **11**, 79.
204. Hotham-Iglewski, B., Phillips, L. A., and Franklin, R. M. (1968). Viral RNA transcription-translation complex in *Escherichia coli* infected with bacteriophage R17. *Nature* **219**, 700.
205. Howatson, A. F., and Whitmore, G. F. (1962). The development and structure of vesicular stomatitis virus. *Virology* **16**, 466.
206. Hoyle, L. (1952). Structure of influenza virus. The relation between biological activity and chemical structure of virus fractions. *J. Hyg.* **50**, 229.
207. Hoyle, L., Horne, R. W., and Waterson, A. P. (1961). The structure and composition of the myxoviruses. II. Components released from the influenza virus particle by ether. *Virology* **13**, 448.
208. Hradecna. Z., and Szybalski, W. (1967). Fractionation of the complementary strands of coliphage λ DNA based on the asymmetric distribution of the poly I, G-binding sites. *Virology* **32**, 633.
209. Huebner, R. J., Rowe, W. P., and Lane, W. T. (1962). Oncogenic effects in hamsters of human adenovirus types 12 and 18. *Proc. Natl. Acad. Sci. U.S.* **48**, 2051.
210. Huebner, R. J., Rowe, W. P., Turner, H. C., and Lane, W. T. (1963). Specific adenovirus complement-fixing antigens in virus-free hamster and rat tumors. *Proc. Natl. Acad. Sci. U.S.* **50**, 379.
211. Hummeler, K., Koprowski, H., and Wiktor, T. J. (1967). Structure and development of rabies virus in tissue culture. *J. Virol.* **1**, 152.
212. Hunt, J. A. (1965). Terminal-sequence studies of high-molecular weight ribonucleic acid. The reaction of periodate oxidized ribonucleosides, 5′ ribonucleotides and ribonucleic acid with isoniazide. *Biochem. J.* **95**, 541.
213. Hutchison, C. A., III, Edgell, M. H., and Sinsheimer, R. L. (1967). The process of infection with bacteriophage øX174. XII. Phenotypic mixing between electrophoretic mutants. *J. Mol. Biol.* **23**, 553.
214. Hyde, J. H. Gafford, L. G., and Randall, C. C. (1967). Molecular weight determination of fowlpox virus DNA by electron microscopy. *Virology* **33**, 112.
215. Igel, H. J., and Black, P. H. (1967). *In vitro* transformation by adenovirus-SV40 hybrid viruses. III. Morphology of tumors induced with transformed cells. *J. Exptl. Med.* **125**, 647.
216. Ikeda, H., and Tomizawa, J.-I. (1965). Transducing fragments in generalized transduction by phage P1. *J. Mol. Biol.* **14**, 85.
216a. Inouye, M., and Tsugita, A. (1968). Amino acid sequence of T2 phage lysozyme. *J. Mol. Biol.* **37**, 213.
217. Inouye, M., Yahata, H., Okada, Y., and Tsugita, A. (1968). Change of tryptophan residue at the 158th position of T4 bacteriophage lysozyme into a tyrosine residue by suppression and spontaneous reversion of an amber mutant. *J. Mol. Biol.* **33**, 957.
218. Isaacs, A. (1963). Interferon. *Advan. Virus Res.* **10**, 1.
219. Israel, J. V., Anderson, T. F., and Levine, M. (1967). *In vitro* morpho-

genesis of phage P22 from head and baseplate parts. *Proc. Natl. Acad. Sci. U.S.* **57**, 284.

220. Ito, Y. (1960). A tumor producing factor extracted by phenol from papillomatosis tissue (Shope) of cottontail rabbits. *Virology* **12**, 596.
221. Ito, Y. (1962). Relationship of components of papilloma virus to papilloma and carcinoma cells. *Cold Spring Harbor Symp. Quant. Biol.* **27**, 387.
222. Jacobson, M. F., and Baltimore, D. (1968). Morphogenesis of poliovirus. I. Association of viral RNA with coat protein. *J. Mol. Biol.* **33**, 369.
223. Jacobson, M. F., and Baltimore, D. (1968). Polypeptide cleavage in the formation of poliovirus proteins. *Proc. Natl. Acad. Sci. U.S.* **61**, 77.
224. Jenner, E. (1798). "An inquiry into the Causes and Effects of the Variolae Vaccina, a Disease Discovered in Some of the Western Counties of England, Particularly Gloucestershire, and Known by the Name of Cowpox." Reprinted by Cassell and Co., Ltd., 1896. Available in pamphlet vol. 4232, Army Medical Library, Washington, D.C.
225. Jockusch, H. (1966). Temperatur-sensitive Mutanten des Tabakmosaikvirus I. *In vivo*-Verhalten. *Z. Vererbungslehre* **98**, 320. (1966) II. *In vitro*-Verhalten. *Z. Vererbungslehre* **98**, 344.
226. Joklik, W. K. (1964). The intracellular uncoating of poxvirus DNA. 2. The molecular basis of the uncoating process. *J. Mol. Biol.* **8**, 277.
227. Joklik, W. K. (1968). The large DNA animal viruses: The poxvirus and herpesvirus group. *In* "Molecular Basis of Virology" (H. Fraenkel-Conrat, ed.), p. 576. Reinhold Publishing Corp., New York.
228. Joklik, W. K., and Becker, Y. (1964). The replication and coating of vaccinia DNA. *J. Mol. Biol.* **10**, 452.
299. Joklik, W. K., Jungwirth, C., Oda, K., and Woodson, B. (1967). The vaccinia virus multiplication cycle. *In* "The Molecular Biology of Viruses" (J. S. Colter and W. Paranchych, eds.), p. 473. Academic Press, New York.
230. Josse, J., Kaiser, A. D., and Kornberg, A. (1961). Enzymatic synthesis of DNA. VIII. Frequencies of nearest neighbor base sequences in DNA. *J. Biol. Chem.* **236**, 864.
231. Kado, C. I., and Knight, C. A. (1966). Location of a local lesion gene in tobacco mosaic virus RNA. *Proc. Natl. Acad. Sci. U.S.* **55**, 1276.
232. Kado, C. I., and Knight, C. A. (1968). The coat protein gene of TMV. I. Location of the gene by mixed infection. *J. Mol. Biol.* **36**, 15.
233. Kaiser, A. D. (1957). Mutations in a temperate bacteriophage affecting its ability to lysogenize *E. Coli*. *Virology* **3**, 42.
234. Kalmakoff, J., and Tremaine, J. H. (1968). Physicochemical properties of tipula iridescent virus. *J. Virol.* **2**, 738.
235. Kanner, L. C., and Kozloff, L. M. (1964). The reaction of indole and T2 bacteriophage. *Biochemistry* **3**, 215.
236. Kaper, J. M. (1968). The small RNA viruses of plants, animals and bacteria. A. Physical properties. *In* "Molecular Basis of Virology" (H. Fraenkel-Conrat, ed.), p. 1. Reinhold Publishing Corp., New York.

237. Kaper, J. M., and Halperin, J. E. (1965). Alkaline degradation of turnip yellow mosaic virus. II. In situ breakage of the ribonucleic acid. *Biochemistry* **4**, 2434.
238. Kaper, J. M., and Jenifer, F. G. (1965). Studies on the interaction of *p*-chloromercuribenzoate with turnip yellow mosaic virus. III. Involvement of the ribonucleic acid. *Arch. Biochem. Biophys.* **112**, 331.
239. Kaper, J. M., and Jenifer, F. G. (1967). Studies on the interaction of *p*-mercuribenzoate with turnip yellow mosaic virus. IV. Conformational phage exposure of buried prototropic groups, and pH-induced degradation. *Biochemistry* **6**, 440.
240. Kass, S. J., and Knight, C. A. (1965). Purification and chemical analysis of Shope papilloma virus. *Virology* **27**, 273.
241. Kassanis, B. (1968). Satellitism and related phenomena in plant and animal viruses. *Advan. Virus Res.* **13**, 147.
242. Kates, M., Allison, A. C., Tyrell, D. A. J., and James, A. T. (1961). Lipids of influenza virus and their relation to those of the host cell. *Biochem. Biophys. Acta* **52**, 455.
243. Kates, M., Allison, A. C., Tyrell, D. A. J., and James, A. T. (1962). Origin of lipids in influenza virus. *Cold Spring Harbor Symp. Quant. Biol.* **27**, 293.
244. Kates, J. R., and McAuslan, B. R. (1967). Pox virus DNA-dependent RNA polymerase. *Proc. Natl. Acad. Sci.* **58**, 134.
245. Kelley, J. J., and Kaesberg, P. (1962). Biophysical and biochemical properties of top component and bottom component of alfalfa mosaic virus. *Biochim. Biophys. Acta* **61**, 865.
246. King, J. (1968). Assembly of the tail of bacteriophage T4. *J. Mol. Biol.* **32**, 231.
247. Kiger, J. A., Young, E. T., and Sinsheimer, R. L. (1967). Infectivity of single stranded rings of bacteriophage λ DNA. *J. Mol. Biol.* **28**, 157.
248. Kiger, J. A., Young, E. T., and Sinsheimer, R. L. (1968). Purification and properties of intracellular λ DNA rings. *J. Mol. Biol.* 33, 395.
249. Kleinschmidt, A. K., Lang, D., Jacherts, D., and Zahn, R. K. (1962). Darstellung des gesamten Deoxyribosenukleinsäure-inhalts von T2 Bakteriophagen. *Biochim. Biophys. Acta* **61**, 857.
250. Klug, A., and Caspar, D. L. D. (1960). The structure of small viruses. *Advan. Virus Res.* **7**, 225.
251. Klug, A., Longley, W., and Leberman, R. (1966). Arrangement of protein subunits and the distribution of nucleic acid in turnip yellow mosaic virus. I. X-ray diffraction studies. *J. Mol. Biol.* **15**, 315.
252. Knight, C. A. (1944). A sedimentable component of allantoic fluid and its relationship to influenza viruses. *J. Exptl. Med.* **80**, 83.
253. Koch, M. A., Eggers, H. J., Anderer, F. A., Schlumberger, H. D., and Frank, H. (1967). Structure of Simian virus 40. I. Purification and physical characterization of the virus particle. *Virology* **32**, 503.
254. Kohn, A. (1965). Polykaryocytosis induced by NDV in monolayers of animal cells. *Virology* **26**, 228.

255. Koprowski, H., Jensen, F. C., and Steplewski, Z. (1967). Activation of production of infectious tumor virus SV40 in heterokaryon cultures. *Proc. Natl. Acad. Sci. U.S.* **58**, 127.
256. Kornberg, A. (1962). "Enzymatic Synthesis of DNA." John Wiley and Sons, New York.
257. Kornberg, A., Zimmerman, S. B., Kornberg, S. R., and Josse, J. (1959). Enzymatic synthesis of deoxyribonucleic acid. VI. Influence of bacteriophage T2 on the synthetic pathway of host cells. *Proc. Natl. Acad. Sci. U.S.* **45**, 772.
258. Kozloff, L. M. (1968). Biochemistry of the T-even bacteriophages of *Escherichia coli*. *In* "Molecular Basis of Virology" (H. Fraenkel-Conrat, ed.), p. 435. Reinhold Publishing Corp., New York.
259. Kramer, E. (1957). Electrophoretische Untersuchungen an Mutanten des Tabakmosaikvirus. *Z. Naturforsch.* **12b**, 609.
260. Lauffer, M. A., and Stevens, C. L. (1968). Structure of the TMV particle; polymerization of TMV protein. *Advan. Virus Res.* **13**, 1.
261. Laver, W. G. (1964). Structural studies on the protein subunits from three strains of influenza virus. *J. Mol. Biol.* **9**, 109.
262. Laver, W. G., and Kilbourne, E. D. (1966). Identification in a recombinant influenza virus of structural proteins derived from both parents. *Virology* **30**, 493.
263. Laver, W. G., Suriano, J. R., and Green, M. (1967). Adenovirus proteins. II. N-terminal amino acid analysis. *J. Virol.* **1**, 723.
264. Lawley, P. D. (1966). Effects of some chemical mutagens and carcinogens on nucleic acids. *In* "Progress in Nucleic Acid Research and Molecular Biology," Vol. **5** (J. N. Davidson and Waldo E. Cohn, eds.), p. 89. Academic Press, New York.
265. Lawley, P. D. (1967). Reaction of hydroxylamine at high concentration with deoxycytidine or with polycytidylic acid: Evidence that substitution of amino groups in cytosine residues by hydroxylamino is a primary reaction and the possible relevance to hydroxylamine mutagenesis. *J. Mol. Biol.* **24**, 75.
266. Lee, P. E. (1968). Partial purification of wheat striate mosaic virus and fine structural studies of the virus. *Virology* **34**, 583.
267. Lennox, E. S. (1955). Transduction of linked genetic characters of the host by bacteriophage P1. *Virology* **1**, 190.
268. Levinthal, C., Hosoda, J., and Shub, D. (1967). The control of protein synthesis after phage infection. *In* "Molecular Biology of Viruses" (J. S. Colter and W. Paranchych, eds.). Academic Press, New York.
269. Lief, F. S., and Henle, W. (1956). Studies on the soluble antigen of influenza virus. 1. The release of S antigen from elementary bodies by treatment with ether. *Virology* **2**, 753.
270. Lin, J.-Y., Tsung, C.-M., and Fraenkel-Conrat, H. (1967). The coat protein of the RNA phage MS2. *J. Mol. Biol.* **24**, 1.
270a. Lin, J.-Y., and Fraenkel-Conrat, H. (1967). Demonstration of further

differences between *in vitro* and *in vivo* synthesized MS2 coat protein. *Biochemistry* **6**, 3402.
271. Lister, R. M. (1966). Possible relationships of virus-specific products of tobacco rattle virus infections. *Virology* **28**, 350.
272. Litman, R. M., and Pardee, A. B. (1956). The production of bacteriophage mutants by a disturbance of DNA metabolism. *Nature* **178**, 529.
273. Lodish, H. F. (1968). Bacteriophage f2 RNA: Control of translation and gene order. *Nature* **220**, 345.
274. Loeffler, F., and Frosch, P. (1898). Berichte der Kommission zur Erforschung der Maul und Klauenseuche bei dem Institut für Infektions-Krankheiten in Berlin. *Zentr. Bakter., Abt. I Orig.* **23**, 371.
275. Loh, P. C., and Shatkin, A. J. (1968). Structural proteins of reoviruses. *J. Virol.* **2**, 1353.
276. Luria, S. E., Williams, R. C., and Backus, R. C. (1951). Electron micrographic counts of bacteriophage particles. *J. Bacteriol.* **61**, 179.
277. Lwoff, A. (1966). *In* Phage and the Origins of Molecular Biology (J. Cairns, G. S. Stent and J. D. Watson, eds.), p. 88. *Cold Spring Harbor Lab. Quant. Biol.,* Cold Spring Harbor, New York.
278. Lwoff, A., Siminovitch, L., and Kjeldgaard, N. (1950). Induction de la lyse bactériophagique de la totalité d'une population microbienne lysogène. *Compt. Rend.* **231**, 190.
279. MacHattie, L. A., and Thomas, C. A., Jr. (1964). DNA from bacteriophage λ: Molecular length and conformation. *Science* **144**, 1142.
280. MacHattie, L. A., Ritchie, D. A., and Thomas, C. A., Jr. (1967). Terminal repetition in permuted T2 bacteriophage DNA molecules. *J. Mol. Biol.* **23**, 355.
281. Macpherson, I., and Montaguier, L. (1964). Agar suspension culture for the selective assay of cells transformed by polyoma virus. *Virology* **23**, 291.
282. Maizel, J. V., Jr. (1963). Evidence for multiple components in the structural protein of type 1 poliovirus. *Biochem. Biophys. Res. Commun.* **13**, 483.
283. Maizel, J. V., Jr., and Summers, D. F. (1968). Evidence for differences in size and composition of the poliovirus-specific polypeptides in infected HeLa cells. *Virology* **36**, 48.
284. Maizel, J. V., Jr., Phillips, B. A., and Summers, D. F. (1967). Composition of artificially produced and naturally occurring empty capsids of poliovirus type 1. *Virology* **32**, 692.
285. Maizel, J. V., Jr., White, D. O., and Scharff, M. D. (1968). The polypeptides of adenovirus. I. Evidence for multiple protein components in the virion and comparison of types 2, 7A, and 12. *Virology* **36**, 115.
286. Maizel, J. V., Jr., White, D. O., and Scharff, M. D. (1968). The polypeptides of adenovirus. II. Soluble proteins, cores, top components and the structure of the virion. *Virology* **36**, 126.
287. Maloney, J. B. (1960). Biological studies on a lymphoid leukemia virus

extracted from sarcoma 37: I. Origin and introductory investigations. *J. Natl. Cancer Inst.* **24**, 933.

288. Markham, R. (1967). The Ultracentrifuge. *In* "Methods in Virology," Vol. II (K. Maramorosch and H. Koprowski, eds.), p. 1. Academic Press, New York.
289. Mandeles, S. (1967). Base sequence at the 5′-linked terminus of TMV-RNA. *J. Biol. Chem.* **242**, 3103.
290. Mandeles, S. (1968). Location of unique sequences in TMV-RNA. *J. Biol. Chem.* **243**, 3671.
291. Marmur, J., Rownd, R., and Schildkraut, C. C. (1963). Denaturation and renaturation of DNA. *Progr. Nucleic Acid. Res.* **7**, 232.
292. Marvin, D. A. (1966). X-ray diffraction and electron microscope studies on the structure of the small filamentous bacteriophage fd. *J. Mol. Biol.* **15**, 8.
293. Matthews, R. E. F. (1960). Properties of nucleoprotein fractions isolated from turnip yellow mosaic virus preparations. *Virology* **12**, 521.
294. Matthews, R. E. F. (1967). Serological techniques for plant viruses. *In* "Methods in Virology," Vol. III (K. Maramorosch and H. Koprowski, eds.), p. 201. Academic Press, New York.
295. May, D. S., and Knight, C. A. (1965). Polar stripping of protein subunits from tobacco mosaic virus. *Virology* **25**, 502.
296. Mayor, H. D., and Jordan, L. E. (1968). Preparation and properties of the internal capsid components of reovirus. *J. Gen. Virol.* **3**, 233.
297. Mayor, H. D., Jamison, R. M., Jordan, L. E., and Mitchell, M. V. (1965). Reoviruses. II. Structure and composition of the virion. *J. Bacteriol.* **89**, 1548.
298. di Mayorca, G. A., Eddy, B. E., Stewart, S. E., Hunter, W. S., Friend, C., and Bendich, A. (1959). Isolation of infectious deoxyribonucleic acid from SE polyoma infected tissue cultures. *Proc. Natl. Acad. Sci. U.S.* **45**, 1805.
299. Mazzone, H. M. (1967). Equilibrium ultracentrifugation. *In* "Methods in Virology," Vol. II (K. Maramorosch and H. Koprowski, eds.), p. 41. Academic Press, New York.
300. McAuslan, B. R. (1963). The induction and repression of thymidine kinase in the poxvirus-infected HeLa cell. *Virology* **21**, 383.
301. McAuslan, B. R., and Kates, J. R. (1967). Poxvirus induced acid DNase: Regulation of synthesis; control of activity *in vivo;* purification and properties of the enzyme. *Virology* **33**, 709.
302. McBride, W. D., and Wiener, A. (1964). *In vitro* transformation of hamster kidney cells by human adenovirus type 12. *Proc. Soc. Exptl. Biol. Med.* **115**, 870.
303. McLaren, A. D., and Shugar, D. (1964). "Photochemistry of Proteins and Nucleic Acids." Pergamon Press; The Macmillan Co., New York.
304. Mellors, R. C. (1960). Tumor cell localization of the antigen of the Shope papilloma and the Rous sarcoma virus. *Cancer Res.* **20**, 749.

305. Meselson, M., and Stahl, F. W. (1958). The replication of DNA in *Escherichia coli. Proc. Natl. Acad. Sci. U.S.*, **44**, 671.
306. Meselson, M., Stahl, F. W., and Vinograd, J. (1957). Equilibrium sedimentation of macromolecules in density gradients. *Proc. Natl. Acad. Sci. U.S.*, **43**, 581.
307. Miki, T., and Knight, C. A. (1965). Preparation of broad bean mottle virus protein. *Virology* **25**, 478.
308. Miki, T., and Knight, C. A. (1968). Protein subunit of potato virus X. *Virology* **36**, 168.
309. Mills, D. R., Peterson, R. L., and Spiegelman, S. (1967). An extracellular Darwinian experiment with a self duplicating nucleic acid molecule. *Proc. Natl. Acad. Sci. U.S.* **58**, 217.
310. Mills, D. R., Bishop, D. H. L., and Spiegelman, S. (1968). The mechanism and direction of RNA synthesis templated by free minus strands of a "little" variant of Qβ RNA. *Proc. Natl. Acad. Sci. U.S.* **60**, 713.
311. Milne, R. G. (1967). Plant viruses inside cells. *Sci. Progr.* (*Oxford*) **55**, 203.
312. Morrison, J. M., Keir, H. M., Subak-Sharpe, H., and Crawford, L. V. (1967). Nearest neighbor base sequence analysis of the deoxyribonucleic acids of a further three mammalian viruses: Simian virus 40, human papilloma virus and adenovirus type 2. *J. Gen. Virol.* **1**, 101.
313. Mudd, O. K., and Marmur, J. (1968). Conversion of bacillus subtilis DNA to phage DNA following mitomycin induction. *J. Mol. Biol.* **34**, 439.
314. Müller, G., Schneider, C. C., and Peters, D. (1966). Zur Feinstruktur des Reovirus (Type 3). *Arch. Ges. Virusforsch.* **19**, 110.
315. Nagington, J., Newton, A. A., and Horne, R. W. (1964). The structure of Orf virus. *Virology* **23**, 461.
316. Nakai, T., and Howatson, A. F. (1968). The fine structure of vesicular stomatitis virus. *Virology* **35**, 268.
317. Narita, K. (1958). Isolation of acetylpeptide from enzymic digests of TMV-protein. *Biochim. Biophys. Acta* **28**, 184.
318. Nathans, D., Notani, G., Schwartz, J. H., and Zinder, N. D. (1962). Biosynthesis of the coat protein of coliphage f2 by *E. coli* extracts. *Proc. Natl. Acad. Sci. U.S.* **48**, 1424.
319. Neurath, A. R. (1964a). Separation of a haemolysin from myxoviruses and its possible relationship to normal chorioallantoic membrane cells. *Acta Virol.* (*Prague*) **8**, 154.
320. Neurath, A. R. (1964b). Association of Sendai virus with esterase and leucine aminopeptidase activity, its probable relationship to haemolysin. *Z. Naturforsch.* **19b**, 810.
321. Newman, I., and Hanawalt, P. (1968). Role of polynucleotide ligase in T4 DNA replication. *J. Mol. Biol.* **35**, 639.
322. Nirenberg, M., and Matthaei, H. (1961). The dependence of cell free protein synthesis in *E. coli* upon naturally occurring or synthetic polyribonucleotides. *Proc. Natl. Acad. Sci. U.S.* **47**, 1588.

323. Nosaka, Y., and Shimizu, I. (1968). Lengths of the nucleocapsids of Newcastle disease and mumps viruses. *J. Mol. Biol.* **35**, 369.
324. Oda, K., and Joklik, W. K. (1967). Hybridization and sedimentation studies on "early" and "late" vaccinia messenger RNA. *J. Mol. Biol.* **27**, 395.
325. Offord, R. E. (1966). Electron microscope observations on the substructure of tobacco rattle virus. *J. Mol. Biol.* **17**, 370.
326. Ogawa, H., and Tomizawa, J. I. (1967). Bacteriophage λ DNA with different structures found in infected cells. *J. Mol. Biol.* **23**, 265.
327. Okamoto, K., Mudel, J. A., Mangan, J., Huang, W. M., Subbaiah, T. V., and Marmur, J. (1968). Properties of the defective phage of bacillus subtilis. *J. Mol. Biol.* **34**, 413.
327a. Okazaki, R., Okazaki, T., Sakabe, K., Sugimoto, K., and Sugino, A. (1968). Mechanism of DNA chain growth. I. Possible discontinuity and unusual secondary structure of newly synthesized chains. *Proc. Natl. Acad. Sci. U.S.* **59**, 598.
328. Olivera, B. M., and Lehmann, I. R. (1967). Linkage of polynucleotides through phosphodiester bonds by an enzyme from *E. coli. Proc. Natl. Acad. Sci. U.S.* **57**, 1426.
329. Orgel, L. E. (1965). The chemical basis of mutation. *Advan. Enzymol.* **27**, 289.
330. Pardee, A. B. (1968). Emphores. *In* "Structural Chemistry and Molecular Biology," p. 216. W. H. Freeman and Co., San Francisco.
331. Passen, S., and Schultz, R. B. (1965). Use of the Shope papilloma virus-induced arginase as a biochemical marker *in vitro*. *Virology* **26**, 122.
332. Pasteur, L. (1885). Méthode pour prévenir la rage après morsure. *Compt. Rend.* **101**, 765.
333. Peacock, A. C., and Dingman, C. W. (1968). Molecular weight estimation and separation of ribonucleic acid by electrophoresis in agarose-acrylamide composite gels. *Biochemistry* **7**, 668.
334. Penman, S., Becker, Y., and Darnell, J. E., Jr. (1964). A cytoplasmic structure involved in the synthesis and assembly of poliovirus components. *J. Mol. Biol.* **8**, 541.
335. Perham, R. N., and Richards, F. M. (1968). Reactivity and structural role of protein amino groups in TMV. *J. Mol. Biol.* **33**, 795.
336. Pettersson, V., Philipson, L., and Höglund, S. (1968). Structural proteins of adenoviruses. II. Purification and characterization of the adenovirus type 2 fiber antigen. *Virology* **35**, 204.
337. Philipson, L., and Choppin, P. W. (1960). On the role of virus sulfhydryl groups in the attachment of entero viruses to erythrocytes. *J. Exptl. Med.* **112**, 455.
338. Philipson, L., Bengtson, S., Brishamm, S., Svennerholm, L., and Zetterquist, O. (1964). Purification and chemical analysis of erythrocyte receptor for haemagglutinating enteroviruses. *Virology* **22**, 580.
339. Piña, M., and Green, M. (1965). Biochemical studies on adenovirus

multiplication. IX. Chemical and base composition analysis of 28 human adenoviruses. *Proc. Natl. Acad. Sci. U.S.* **54**, 547.

340. Plagemann, P. G., and Swim, H. E. (1966). Replication of Mengo virus. 1. Effect on synthesis of macromolecules by host cell. *J. Bacteriol.* **91**, 2317.
341. Poljak, R. J. (1968). A study of the coat proteins of bacteriophage ϕX174. *Virology* **35**, 185.
342. Polson, A., and Russell, B. (1967). Electrophoresis of viruses. *In* "Methods in Virology," Vol. II (K. Maramorosch and H. Koprowski, eds.), p. 39. Academic Press, New York.
343. Pons, M. W., and Hirst, G. K. (1968). Polyacrylamide gel electrophoresis of influenza virus RNA. *Virology* **34**, 385.
344. Ptashne, M. (1967a). Isolation of the λ repressor. *Proc. Natl. Acad. Sci. U.S.* **57**, 306.
345. Ptashne, M. (1967b). The λ phage repressor binds specifically to λ DNA. *Nature* **214**, 232.
346. RajBhandary, U. L. (1968). The labeling of end groups in polynucleotides. The selective modification of diol groups in RNA. *J. Biol. Chem.* **243**, 556.
347. Rappaport, I. (1965). The antigenic structure of TMV. *Advan. Virus Res.* **11**, 223.
348. Rauscher, F. J. (1962). A virus induced disease of mice characterized by erythrocytopoiesis and lymphoid leukemia. *J. Natl. Cancer Inst.* **29**, 515.
349. Reichmann, M. E. (1959a). Potato virus X. II. Preparation and properties of purified, non-aggregated virus from tobacco. *Can. J. Chem.* **37**, 4.
350. Reichmann, M. E. (1959b). Potato virus X. III. Light scattering studies. *Can. J. Chem.* **37**, 384.
351. Reichmann, M. E. (1964). The satellite tobacco necrosis virus: A single protein and its genetic code. *Proc. Natl. Acad. Sci. U.S.* **52**, 1009.
352. Rentschler, L. (1967). Aminosäuresequenzen und physikochemisches Verhalten des Hüllproteins eines Wildstammes des Tabakmosaikvirus. Teil I. Analyse der Primärstruktur des Hüllproteins vom Wildstamm U2. Teil II. Aggregationverhalten und Ladungsverteilung im Vergleich zu den Stämmen vulgare und dahlemense. *Mol. Gen. Genetics* **100**, 84, 96.

352a. Rhoades, M., MacHattie, L. A., and Thomas, C. A., Jr. (1968). The P22 bacteriophage DNA molecule. *J. Mol. Biol.* **37**, 21; 41.

353. Richardson, C. C. (1965). Phosphorylation of nucleic acid by an enzyme from T4 bacteriophage-infected *Escherichia coli*. *Proc. Natl. Acad. Sci. U.S.* **54**, 158.
354. Ritchie, D. A., Thomas, C. A., Jr., MacHattie, L. A., and Wensink, P. C. (1967). Terminal repetition in non-permuted T3 and T7 bacteriophage DNA molecules. *J. Mol. Biol.* **23**, 365.

355. Roberts, J. W., and Steitz, J. E. (Argetsinger) (1967). The reconstitution of infective bacteriophage R17. *Proc. Natl. Acad. Sci. U.S.* **58**, 1416.

356. Robertson, H. D., and Zinder, N. D. (1968). Identification of the terminus of nascent f2 bacteriophage RNA. *Nature* **220**, 69.

357. Robertson, H. D., Webster, R. E., and Zinder, N. D. (1968). Bacteriophage coat protein as repressor. *Nature* **218**, 533.

358. Robinson, W. S., and Duesberg, P. H. (1968). *In* "Molecular Basis of Virology," (H. Fraenkel-Conrat, ed.), p. 255. Reinhold Publishing Corp., New York.

358a. Roblin, R. (1968). The 5′-terminus of bacteriophage R17 RNA:pppGp. *J. Mol. Biol.* **31**, 51.

359. Rombauts, W. A. (1966). The nonacetylated N-terminal sequence of the green tomato atypical mosaic virus coat protein (G-TAMV). *Biochem. Biophys. Res. Commun.* **23**, 549.

360. Rombauts, W. A., and Fraenkel-Conrat, H. (1968). Artificial histidine containing mutants of tobacco mosaic virus. *Biochemistry* **7**, 3334.

361. Rossomando, E. F., and Zinder, N. D. (1968). Studies on bacteriophage f1. I. Alkali-induced disassembly of the phage into DNA and protein. *J. Mol. Biol.* **36**, 387.

362. Rothman, J. (1965). Transduction studies on the relationship between prophage and host chromosome. *J. Mol. Biol.* **12**, 892.

363. Rott, R. (1964). Antigenicity of Newcastle disease virus. *In* "Newcastle Disease Virus, an Evolving Pathogen" (R. P. Hanson, ed.), p. 133. Univ. of Wisconsin Press, Madison.

364. Rous, P. (1911). A sarcoma of the fowl transmissible by an agent separable from the tumor cells. *J. Exptl. Med.* **13**, 397.

365. Rowe, W. P. (1967). Some interaction of defective animal viruses. *In* "Perspectives in Virology" Vol. V, p. 123. Academic Press, New York.

366. Rowe, W. P., and Puch, W. E. (1966). Studies of adenovirus-SV40 hybrid viruses V. Evidence for linkage between their genetic materials. *Proc. Natl. Acad. Sci. U.S.* **55**, 1126.

367. Roy, D., Lesnaw, J., Fraenkel-Conrat, H., and Reichmann, M. E. (1969). The protein subunit of the satellite tobacco necrosis virus. *Virology* **38**, 368.

368. Rubenstein, I., Thomas, C. A., Jr., and Hershey, A. D. (1961). The molecular weights of T2 bacteriophage and its first and second breakage product. *Proc. Natl. Acad. Sci. U.S.* **47**, 1113.

369. Rubin, H. (1962). The immunological basis for non-infective Rous sarcomas. *Cold Spring Harbor Symp. Quant. Biol.* **27**, 441.

370. Rubin, H. (1966). Quantitative tumor viorology. *In* "Phage and the Origins of Molecular Biology" (J. Cairns, G. S. Stent, and J. D. Watson, eds.), p. 292. *Cold Spring Harbor Lab. Quant. Biol.*, Cold Spring Harbor, New York.

371. Rubin, H., and Vogt, P. K. (1962). An avian leukosis virus associated with stocks of Rous sarcoma virus. *Virology* **17**, 184.

372. Rubin, H., and Colby, C. (1968). Early release of growth inhibition in

cells infected with Rous sarcoma virus. *Proc. Natl. Acad. Sci. U.S.* **60**, 482.

373. Rueckert, R. R. (1965). Studies on the structure of viruses of the Columbia SK group. II. The protein subunits of ME virus and other members of the group. *Virology* **26**, 345.
374. Rueckert, R. R., and Duesberg, P. H. (1966). Non-identical peptide chains in mouse encephalitis virus. *J. Mol. Biol.* **17**, 490.
375. Rushizky, G. W., Sober, H. A., and Knight, C. A. (1962). Products obtained by digestion of the nucleic acids of some strains of tobacco mosaic virus with ribonuclease T-1. *Biochim. Biophys. Acta* **61**, 56.
376. Russell, W. C., and Becker, Y. (1968). A maturation factor for adenovirus. *Virology* **35**, 18.
377. Russell, W. C., Laver, W. G., and Sanderson, P. J. (1968). Internal components of adenovirus. *Nature* **219**, 1127.
378. Sadron, C. L. (1960). DNAs as macromolecules. *In* "The Nucleic Acids," Vol. III (E. Chargaff and J. N. Davidson, eds.), p. 1. Academic Press, New York.
379. Sambrook, J., Westphal, H., Srinivasan, P. R., and Dulbecco, R. (1968). The integrated state of viral DNA in SV40-transformed cells. *Proc. Natl. Acad. Sci. U.S.* **60**, 1288.
380. Sanders, F. K. (1964). The infective nucleic acids of animal viruses. *In* "Techniques in Experimental Virology" (R. J. C. Harris, ed.), p. 277. Academic Press, New York.
381. Sänger, H. L. (1968). Characteristics of tobacco rattle virus. I. Evidence that its two particles are functionally defective and mutually complementing. *Mol. Gen. Genetics* **101**, 346.

381a. Sänger, H. L., and Knight, C. A. (1963). Action of actinomycin D on RNA synthesis in healthy and virus-infected tobacco leaves. *Biochem. Biophys. Res. Commun.* **13**, 455.

382. Sarkar, S. (1960). Interaction and mixed aggregation of proteins from tobacco mosaic virus strains. *Z. Naturforsch.* **15b**, 778.
383. Schachman, H. K. (1959). "Ultracentrifugation in Biochemistry." Academic Press, New York.
384. Schachman, H. K., and Williams, R. C. (1959). The physical properties of infectious particles. *In* "The Viruses," Vol. I (F. M. Burnet and W. M. Stanley, eds.), p. 223. Academic Press, New York.
385. Schaffer, F. L., and Schwerdt, C. E. (1955). Crystallization of purified MEF-1 poliomyelitis virus particles. *Proc. Natl. Acad. Sci. U.S.* **41**, 1020.
386. Schlegel, D. E., and Smith, S. (1966). Sites of virus synthesis in plant cells. *In* "Viruses of Plants" (A. B. R. Beemster and J. Dijkstra, eds.), p. 54. Interscience, New York.
387. Schlesinger, M. (1934). Zur Frage der chemischen Zusammensetzung des Bakteriophagen. *Biochem. Z.* **237**, 306.
388. Schlumberger, H. D., Anderer, F. A., and Koch, M. A. (1968). Structure of simian virus 40. IV. The polypeptide chains of the virus particle. *Virology* **36**, 42.

389. Schneider, I. R., Diener, T. O., and Safferman, R. S. (1964). Blue-green algae virus LPP-1. Purification and partial characterization. *Science* **144**, 1127.

390. Schramm, G., and Zillig, W. (1955). Über die Struktur des Tabakmosaikvirus. IV. Die Reaggregation des nucleinsäurefreien Proteins. *Z. Naturforsch.* **10b**, 493.

391. Schramm, G., Schumacher, G., and Zillig, W. (1955). Uber die Struktur des Tabakmosaikvirus. III. Der Zerfall in alkalischer Losung. *Z. Naturforsch.* **10b**, 481.

392. Schroeder, Walter A. (1968). "The Primary Structure of Proteins; Principles and Practices for the Determination of Amino Acid Sequence." Harper and Row, New York.

393. Schuster, H., and Schramm, G. (1958). Bestimmung der biologisch wirksamen Einheit in der Ribosenucleinsäure des Tabakmosaikvirus auf chemischem Wege. *Z. Naturforsch.* **13b**, 697.

394. Schuster, H., and Wittmann, H. G. (1963). The inactivating and mutagenic action of hydroxylamine on tobacco mosaic virus ribonucleic acid. *Virology* **18**, 421.

395. Schuster, H., and Wilhelm, R. C. (1963). Reaction differences between tobacco mosaic virus and its free ribonucleic acid with nitrous acid. *Biochim. Biophys. Acta* **68**, 554.

396. Schwartz, J. H. (1967). Initiation of protein synthesis under the direction of TMV RNA in cell free extracts of *E. coli*. *J. Mol. Biol.* **30**, 309.

397. Scott, D. W. (1965). Serological cross reactions among the RNA-containing coliphages. *Virology* **26**, 85.

398. Sengbusch, P. v. (1965). Aminosäureaustausche und Tertiärstruktur eines Proteins. Vergleich von Mutanten des Tabakmosaikvirus mit serologischen und physiko-chemischen Methoden. *Z. Vererbungslehre* **96**, 364.

399. Setlow, R. B. (1968). Photochemistry, photobiology and repair of polynucleotides. *Progr. Nucleic Res. Mol. Biol.* **8**, 257.

400. Seto, F. T., Drzeniek, R., and Rott, R. (1966). Isolation of a low molecular weight sialidase (neuraminidase) from influenza virus. *Biochim. Biophys. Acta* **113**, 402.

401. Shapiro, A. L., Viñuela, E., and Maizel, J. V., Jr. (1967). Molecular weight estimation of polypeptide chains by electrophoresis in SDS-polyacrylamide gels. *Biochim. Biophys. Res. Commun.* **28**, 815.

402. Shapiro, L., Franze de Fernandez, M. T., and August, J. T. (1968). Resolution of two factors required in the QβRNA polymerisation reaction. *Nature* **220**, 478.

403. Shapiro, R. (1964). Isolation of a 2-nitropurine from the reaction of guanosine with nitrous acid. *J. Am. Chem. Soc.* **86**, 2948.

404. Shatkin, A. J. (1968). Viruses containing double stranded RNA. *In* "Molecular Basis of Virology" (H. Fraenkel-Conrat, ed.), p. 351. Reinhold Publishing Corp., New York. See also (1969). *Advan. Virus Res.* **14**, 63.

405. Shatkin, A. J., Sipe, J. D., and Loh, P. C. (1968). Separation of ten reovirus genome segments by polyacrylamide gel electrophoresis. *J. Virology* **2**, 986.

406. Shaw, J. G. (1967). *In vivo* removal of protein from TMV after inoculation of tobacco leaves. *Virology* **31**, 665.

406a. Shedlovsky, A., and Brenner, S. (1963). A chemical basis for the host-induced modification of T-even bacteriophages. *Proc. Natl. Acad. Sci. U.S.* **50**, 300.

407. Shepherd, R. J., Wakeman, R. J., and Romanko, R. R. (1968). DNA in cauliflower mosaic virus. *Virology* **36**, 150.

408. Shimura, Y., Moses, R. E., and Nathans, D. (1967). Coliphage MS2 containing 5-fluorouracil. II. RNA deficient particles formed in the presence of 5-fluorouracil. *J. Mol. Biol.* **28**, 95.

409. Shimura, Y., Kaizer, H., and Nathans, D. (1968). Fragments of MS2 RNA as messengers for specific bacteriophage proteins: fragments from fluorouracil-containing particles. *J. Mol. Biol.* **38**, 453.

410. Shope, R. E. (1933). Infectious papillomatosis of rabbits. *J. Exptl. Med.* **58**, 607.

411. Sia, C. L., and Horecker, B. L. (1968). Dissociation of protein subunits by maleylation. *Biochem. Biophys. Res. Commun.* **31**, 731.

412. Siegel, A., Ginoza, W., and Wildman, S. G. (1957). The early events of infection with TMV-RNA. *Virology* **3**, 554.

413. Siegel, A., Zaitlin, M., and Sehgal, O. P. (1962). The isolation of defective tobacco mosaic virus strains. *Proc. Natl. Acad. Sci. U.S.* **48**, 1845.

414. Siegel, A., Hills, G. J., and Markham, R. (1966). *In vitro* and *in vivo* aggregation of the defective PM2 TMV protein. *J. Mol. Biol.* **19**, 140.

415. Silverstein, S. C., and Marcus, P. I. (1964). Early stages of NDV-HeLa cell interaction. *Virology* **23**, 370.

416. Simon, L. D., and Anderson, T. F. (1967). The infection of *Escherichia coli* by T2 and T4 bacteriophages as seen in the electron microscope. II. Structure and function of the baseplate. *Virology* **32**, 298.

417. Simon, L. D., and Anderson, T. F. (1967). The infection of *Escherichia coli* by T2 and T4 bacteriophages as seen in the electron microscope. I. Attachment and penetration. *Virology* **32**, 279.

418. Simpson, R. S., and Hauser, R. E. (1966). Structural components of vesicular stomatis virus. *Virology* **29**, 654.

419. Singer, B., and Fraenkel-Conrat, H. (1961). Effects of bentonite on infectivity and stability of TMV-RNA. *Virology* **14**, 59.

420. Singer, B., and Fraenkel-Conrat, H. (1967). Chemical modification of viral RNA. VI. The action of *N*-methyl-*N*′-nitro-*N*-nitrosoguanidine. *Proc. Natl. Acad. Sci. U.S.* **58**, 234.

421. Singer, B., and Fraenkel-Conrat, H. (1969). The role of conformation in chemical mutagenesis. *Progr. Nucleic Acid Res. Mol. Biol.* **9**, 1.

422. Singer, B., Fraenkel-Conrat, H., Greenberg, J., and Michelson, A. M. (1968). Reaction of nitrosoguanidine (*N*-methyl-*N*′-nitro-*N*-nitrosoguanidine) with tobacco mosaic virus and its RNA. *Science* **160**, 1235.

423. Sinha, N. K., Enger, M. D., and Kaesberg, P. (1965). Comparison of pancreatic RNase digestion products of R17 viral RNA and M12 viral RNA. *J. Mol. Biol.* **12**, 299.
424. Sinsheimer, R. L. (1959). A single-stranded deoxyribonucleic acid from bacteriophage ϕX174. *J. Mol. Biol.* **1**, 43.
425. Sinsheimer, R. L. (1968). Bacteriophage ϕX174 and related viruses. *Progr. Nucleic Acid Res. Mol. Biol.* **8**, 115.
426. Sinsheimer, R. L. (1968). *Cold Spring Harbor Symp. Quant. Biol.* 33.
427. Smith, I. O., Gehle, W. D., and Thiel, J. R. (1966). Properties of a small virus associated with adenovirus type 4. *J. Immunol.* **97**, 754.
428. Spiegelman, S., Haruna, I., Holland, I. B., Beaudreau, G., and Mills, D. R. (1965). The synthesis of a self-propagating and infectious nucleic acid with a purified enzyme. *Proc. Natl. Acad. Sci. U.S.* **54**, 919.
429. Spirin, A. S. (1963). Some problems concerning the macromolecular structure of RNA. *Progr. Nucleic Acid Res. Mol. Biol.* **1**, 301.
430. Staehelin, M. (1958). Reaction of tobacco mosaic virus nucleic acid with formaldehyde. *Biochim. Biophys. Acta* **29**, 410.
431. Stanley, W. M. (1935). Isolation of a crystalline protein possessing the properties of the tobacco mosaic virus. *Science* **81**, 644.
432. Stanley, W. M. (1936). The inactivation of crystalline tobacco mosaic virus protein. *Science* **83**, 626.
433. Steere, R. L. (1959). The purification of plant viruses. *Advan. Virus Res.* **6**, 1.
434. Steere, R. L. (1963). Tobacco mosaic virus: Purifying and sorting associated particles according to length. *Science* **140**, 1089.
435. Steinschneider, A., and Fraenkel-Conrat, H. (1966). Studies of nucleotide sequences in tobacco mosaic virus ribonucleic acid III. Periodate oxidation and semicarbazone formation. *Biochemistry* **5**, 2729.
436. Steinschneider, A., and Fraenkel-Conrat, H. (1966). Studies of nucleotide sequences in tobacco mosaic virus ribonucleic acid IV. Use of aniline in step-wise degradation. *Biochemistry* **5**, 2735.
437. Steitz, J. A. (1968). Identification of the A protein as a structural component of bacteriophage R17. *J. Mol. Biol.* **33**, 923.
438. Steitz, J. A. (1968). Isolation of the A protein from bacteriophage R17. *J. Mol. Biol.* **33**, 937.
439. Stent, G. S. (1963). "Molecular Biology of Bacterial Viruses." W. H. Freeman & Co., San Francisco.
440. Stewart, S. E., and Eddy, B. E. (1959). Properties of a tumor-inducing virus recovered from mouse neoplasms. *Perspectives Virol.* **1**, 245.
441. Strauss, J. H., Jr., Burge, B. W., Pfefferkorn, E. R., and Darnell, J. E., Jr. (1968). Identification of the membrane protein and "core" protein of Sindbis virus. *Proc. Natl. Acad. Sci. U.S.* **59**, 533.
442. Streeter, D. G., and Gordon, M. P. (1968). Studies of the role of the coat protein in the U.V. photoinactivation of U1 and U2 strains of TMV. *Photochem. Photobiol.* **8**, 81.
443. Streisinger, G., Edgar, R. S., and Denhardt, G. H. (1964). Chromosome

structure in phage T4. I. Circularity of the linkage map. *Proc. Natl. Acad. Sci. U.S.* **51**, 775.

444. Stretton, A. O. W., Kaplan, S., and Brenner, S. (1966). Nonsense codons. *Cold Spring Harbor Symp. Quant. Biol.* **31**, 173.
445. Subak-Sharpe, H., Bürk, R. R., Crawford, L. V., Morrison, J. M., Hay, J., and Keir, H. M. (1966). An approach to evolutionary relationships of mammalian DNA viruses through analysis of the pattern of nearest neighbor base sequences. *Cold Spring Harbor Symp. Quant. Biol.* **31**, 737.
446. Sueoka, N., Marmur, J., and Doty, P. (1959). Dependence of the density of DNA on guanine-cytosine content. *Nature* **183**, 1429.
447. Sugiyama, T. (1965). 5′ linked end group of RNA from bacteria phage MS2, *J. Mol. Biol.* **11**, 856.
448. Sugiyama, T. (1966). Tobacco mosaic virus-like rods formed by "mixed reconstitution" between MS2 ribonucleic acid and tobacco mosaic virus protein. *Virology* **28**, 488.
449. Sugiyama, T., and Fraenkel-Conrat, H. (1961). Identification of 5′-linked adenosine as end group of TMV-RNA. *Proc. Natl. Acad. Sci. U.S.* **47**, 1393.
450. Sugiyama, T., and Fraenkel-Conrat, H. (1962). The end groups of tobacco mosaic virus RNA. II. Nature of the 3′-linked chain end in TMV and of both ends in four strains. *Biochemistry* **2**, 332.
451. Sugiyama, T., and Nakada, D. (1967). Control of translation of MS2 RNA cistrons by MS2 coat protein. *Proc. Natl. Acad. Sci. U.S.* **57**, 1744.
452. Sugiyama, T., and Nakada, D. (1968). Translational control of bacteriophage MS2 RNA cistrons by MS2 coat protein: polyacrylamide gel electrophoretic analysis of proteins synthesized *in vitro*. *J. Mol. Biol.* **31**, 431.
453. Sugiyama, T., Herbert, R. R., and Hartman, K. A. (1967). Ribonucleoprotein complexes formed between MS2-RNA and MS2 protein *in vitro*. *J. Mol. Biol.* **25**, 455.
454. Summers, W. C. (1968). Equal transfer of both parental T7 DNA strands to progeny bacteriophage. *Nature* **219**, 159.
455. Summers, D. F., Maizel, J. V., Jr., and Darnell, J. E., Jr. (1965). Evidence for virus-specific non-capsid proteins in poliovirus-infected HeLa cells. *Proc. Natl. Acad. Sci. U.S.* **54**, 505.
456. Suzuki, J., and Haselkorn, R. (1968). Studies on the 5′ terminus of TYMV-RNA. *J. Mol. Biol.* **36**, 47.
457. Takahashi, W. N., and Ishii, M. (1952). The formation of rodshaped particles resembling TMV by polymerisation of a protein from mosaic-diseased tobacco leaves. *Phytopathol.* **42**, 690.
458. Taylor, K., Hradecna, Z., and Szybalski, W. (1967). Asymmetric distribution of transcribing regions on the complementary strands of coliphage λ DNA. *Proc. Natl. Acad. Sci. U.S.* **57**, 1618.

459. Temin, H. M. (1964). The participation of DNA in Rous sarcoma virus production. *Virology* **23**, 486.
460. Temin, H., and Rubin, H. (1958). Characteristics of an assay for Rous sarcoma virus and Rous sarcoma cells in tissue culture. *Virology* **6**, 669.
461. Terzaghi, E., Okada, Y., Streisinger, G., Emrich, J., Inouyi, M., and Tsugita, A. (1966). Change of a sequence of amino acids in phage T4 lysozyme by acridine-induced mutations. *Proc. Natl. Acad. Sci. U.S.* **56**, 500.
462. Thirion, J. P., and Kaesberg, P. (1968). The pyrimidine catalogs of M12 and R17 RNAs. *J. Mol. Biol.* **33**, 379.
463. Thirion, J. P., and Kaesberg, P. (1968). Sequence determination of oligonucleotides obtained from pancreatic ribonuclease digests of M12 and R17-RNA. *Biochim. Biophys. Acta* **161**, 297.
464. Thorne, H. V., and Warden, D. (1967). Electrophoretic evidence for a single protein component in the capsid of polyoma virus. *J. Gen. Virol.* **1**, 135.
465. Thorne, H. V., Joyce, E., and Warden, D. (1968). Detection of biologically defective molecules in component I of polyoma virus DNA. *Nature* **219**, 728.
466. Todaro, G. J., and Baron, S. (1965). The role of interferon in the inhibition of SV40 transformation of mouse cell line, 3T3. *Proc. Natl. Acad. Sci. U.S.* **54**, 752.
467. Tomlinson, R. V., and Tener, G. M. (1962). The use of urea to eliminate the secondary binding forces in ion exchange chromatography of polynucleotides. *J. Am. Chem. Soc.* **84**, 2644.
468. Tooze, J., and Weber, K. (1967). Isolation and characterization of amber mutants of bacteriophage R17. *J. Mol. Biol.* **28**, 311.
469. Trager, G. W., and Rubin, H. (1966). Mixed clones produced following infection of chick embryo cultures with Rous sarcoma virus. *Virology* **30**, 275.
470. Traub, P., and Nomura, M. (1968). Structure and function of *Escherichia coli* ribosomes. I. Partial fractionation of the functionally active ribosomal proteins and reconstitution of artificial subribosomal particles. *J. Mol. Biol.* **34**, 575.
471. Trenkner, E., Bonhoeffer, F., and Gierer, A. (1967). The fate of the protein component of bacteriophage f1 during infection. *Biochem. Biophys. Res. Commun.* **28**, 932.
472. Thentin, J. J., Yabe, Y., and Taylor, G. (1962). The quest for human cancer viruses. *Science* **137**, 835.
473. Tsugita, A., and Fraenkel-Conrat, H. (1962). The composition of proteins of chemically evoked mutants of TMV-RNA. *J. Mol. Biol.* **4**, 73.
473a. Tsugita, A., and Inouye, M. (1968). Complete primary structure of phage lysozyme. *J. Mol. Biol.* **37**, 201.
474. Tsugita, A., Gish, D. T., Young, J., Fraenkel-Conrat, H., Knight, C. A., and Stanley, W. M. (1960). The complete amino acid sequence of the protein of tobacco mosaic virus. *Proc. Natl. Acad. Sci. U.S.* **46**, 1463.

475. Twort, F. W. (1915). An investigation on the nature of ultramicroscopic viruses. *Lancet* **189,** (2) 1241.
476. Valentine, R. C., and Pereira, H. G. (1965). Antigens and structure of the adenovirus. *J. Mol. Biol.* **13,** 13.
477. Valentine, R. C., and Strand, M. (1965). Complexes of F-pili and RNA bacteriophage. *Science* **148,** 511.
478. Valentine, R. C., and Wedel, H. (1965). The extracellular stages of RNA bacteriophage infection. *Biochem. Biophys. Res. Commun.* **21,** 106.
479. Valentine, R. C., Wedel, H., and Ippen, K. A. (1965). F-pili requirement for RNA bacteriophage adsorption. *Biochem. Biophys. Res. Commun.* **21,** 277.
480. van Kammen, A. (1968). The relationship between the components of cowpea mosaic virus. I. Two ribonucleoprotein particles necessary for the infectivity of CPMV. *Virology* **34,** 312.
481. van Ravenswaay Claasen, J. C., van Leeuwen, A. B. J., Duijts, G. A. H., and Bosch, L. (1967). *In vitro* translation of alfalfa mosaic virus RNA. *J. Mol. Biol.* **23,** 535.
482. van Regenmortel, M. H. V. (1964). Purification of plant viruses by zone electrophoresis. *Virology* **23,** 495.
483. van Regenmortel, M. H. V. (1967). Serological studies on naturally occurring strains and chemically induced mutants of TMV. *Virology* **31,** 467.
484. van Vloten-Doting, L., Kruseman, J., and Jaspars, E. M. J. (1968). The biological function and mutual dependence of bottom component and top components of alfalfa mosaic virus. *Virology* **34,** 728.
485. Vasquez, C., and Kleinschmidt, A. K. (1968). Electronmicroscopy of RNA strands released from individual reovirus particles. *J. Mol. Biol.* **34,** 137.
486. Vinograd, J., Lebowitz, J., Radloff, R., and Laipis, P. (1965). The twisted circular form of polyoma viral DNA. *Proc. Natl. Acad. Sci. U.S.* **53,** 1104.
487. Vinson, C. G. (1927). Precipitation of the virus of tobacco mosaic. *Science* **66,** 357.
488. Vinson, C. G., and Petre, A. M. (1929). Mosaic disease of tobacco. *Botan. Gaz.* **87,** 14.
489. Viñuela, E., Algranati, I. C., and Ochoa, S. (1967). Synthesis of virus-specific proteins in *Escherichia coli* infected with the RNA bacteriophage MS2. *European J. Biochem.* **1,** 3.
490. Vogt, P. K. (1967). A virus released by "nonproducing" Rous sarcoma cells. *Proc. Natl. Acad. Sci. U.S.* **58,** 801.
491. Vogt, P. K. (1967). Virus-directed host responses in the avian leucosis and sarcoma complex. *Perspectives Virol.* **5,** 199.
492. Vogt, M., and Dulbecco, R. (1963). Steps in neoplastic transformation of hamster embryo cells by polyoma virus. *Proc. Natl. Acad. Sci. U.S.* **49,** 171.

493. Volkin, E., and Astrachan, L. (1956). Phosphorus incorporation in *Escherichia coli* ribonucleic acid after infection with bacteriophage T2. *Virology* **2**, 149.
494. de Wachter, R., and Fiers, W. (1967). Studies on the bacteriophage MS2. IV. The 3′-OH terminal undecanucleotide sequence of the viral RNA chain. *J. Mol. Biol.* **30**, 507.
495. Wagner, G. W., and Bancroft, J. B. (1968). The self assembly of spherical viruses with mixed coat proteins. *Virology* **34**, 748.
496. Waterson, A. P., Hurrell, J. M. W., and Jensen, K. E. (1963). The fine structure of influenza A, B, and C viruses. *Arch. Ges. Virusforsch* **12**, 487.
497. Watkins, J. F., and Dulbecco, R. (1967). Production of SV40 virus in heterokaryons of transformed and susceptible cells. *Proc. Natl. Acad. Sci. U.S.* **58**, 1396.
498. Watson, J. D., and Crick, F. H. C. (1953). (a) A structure for deoxyribose nucleic acids. *Nature* **171**, 737. (b) Genetical implication of the structure of deoxyribose nucleic acid. *Nature* **171**, 964.
499. Watson, D. H., and Wildy, P. (1963). Some serological properties of herpes virus particles studied with the electronmicroscope. *Virology* **21**, 100.
500. Weber, K., and Konigsberg, W. (1967). Amino acid sequence of the f2 coat protein. *J. Biol. Chem.* **242**, 3563.
501. Webster, R. G., and Laver, W. G. (1967). Preparation and properties of antibody directed specifically against the neuraminidase of influenza virus. *J. Immunol.* **99**, 49.
502. Webster, R. G., Laver, W. G., and Kilbourne, E. D. (1968). Reactions of surface antigens of antibodies with influenza virus. *J. Gen. Virol.* **3**, 315.
503. Wecker, E. (1958). The extraction of infectious virus nucleic acid with hot phenol. *Virology* **7**, 241.
504. Weigle, J. (1966). Assembly of phage λ *in vitro*. *Proc. Natl. Acad. Sci. U.S.* **55**, 1462.
505. Weigle, J. (1968). Studies on head-tail union of bacteriophage λ. *J. Mol. Biol.* **33**, 483.
506. Weil, R., and Vinograd, J. (1963). The cyclic helix and cyclic coil forms of polyoma viral DNA. *Proc. Natl. Acad. Sci. U.S.* **50**, 730.
507. Weiss, B., and Richardson, C. C. (1967). The 5′ terminal dinucleotides of the separated strands of T7 bacteriophage DNA. *J. Mol. Biol.* **23**, 405.
508. Weissmann, C., Feix, G., Slor, H., and Pollet, R. (1967). Replication of viral RNA. XIV. Single stranded minus strands as template for the synthesis of viral plus strands *in vitro*. *Proc. Natl. Acad. Sci. U.S.* **57**, 1870.
509. Weith, H. L., Asteriadis, G. T., and Gilham, P. T. (1968). Comparison of RNA terminal sequences of phages f2 and Qβ: Chemical and sedimentation equilibrium studies. *Science* **160**, 1959.

510. Westwood, J. C. N., Harris, W. J., Zwartouw, H. T., Titmuss, D. H. J., and Appleyard, G. (1964). Studies on the structure of vaccinia virus. *J. Gen. Microbiol.* **34**, 67.
511. Wildman, S. G. (1959). The process of infection and virus synthesis with tobacco mosaic virus and other plant viruses. *In* "The Viruses," Vol. 2 (F. M. Burnet and W. M. Stanley, eds.), p. 1. Academic Press, New York.
512. Wildy, P., and Horne, R. W. (1963). Structure of animal virus particles, 1. *In* "Progress in Medical Virology," Vol. V (E. Berger and J. L. Melnick, eds.), p. 1. S. Karger, New York.
513. Williams, R. C., and Smith, K. M. (1958). The polyhedral form of the tipula iridescent virus. *Biochim. Biophys. Acta* **28**, 464.
514. Williams, J. F., and Till, J. E. (1964). Transformation of rat embryo cells in culture by polyoma virus. *Virology* **24**, 505.
515. Williams, R. C. (1953). A method of freeze-drying for electron microscopy. *Exptl. Cell Res.* **4**, 188.
516. Williams, R. C., Backus, R. C., and Steere, R. L. (1951). Macromolecular weights determined by direct particle counting. II. The weight of the tobacco mosaic virus particle. *J. Am. Chem. Soc.* **73**, 2062.
517. Wimmer, E., and Reichmann, M. E. (1968). Pyrophosphate in the 5′ terminal position of a viral RNA. *Science* **160**, 1452.
518. Winocour, E. (1965). Attempts to detect an integrated polyoma genome by nucleic acid hybridization. I. "Reconstruction" experiments and complementarity tests between synthetic polyoma RNA and polyoma tumor DNA. *Virology* **25**, 276.
519. Winocour, E. (1965). Attempts to detect an integrated polyoma genome by nucleic acid hybridization. II. Complementarity between polyoma virus DNA and normal mouse synthetic RNA. *Virology* **27**, 520.
520. Winocour, E. (1968). Further studies on the incorporation of cell DNA into polyoma-related particles. *Virology* **34**, 571. (1969). *Advan. Virus Res.* **14**, 153.
521. Wittmann-Liebold, B. (1966). Aminosäuresequenz im Hüllenprotein des RNS-bacteriophagen fr. *Z. Naturforsch.* **21b**, 1249.
522. Wittmann-Liebold, B., and Wittmann, H. G. (1963). Die primäre Proteinstruktur von Stämmen des Tabakmosaikvirus. Aminosäuresequenzen des Proteins des Tabakmosaikvirusstammes Dahlemense. Teil III: Diskussion der Ergebnisse. *Z. Vererbungslehre* **94**, 427.
523. Wittmann-Liebold, B., and Wittmann, H. G. (1965). Lokaliserung von Aminosäureaustauschen bei Nitritmutanten des Tabakmosaikvirus. *Z. Vererbungslehre* **97**, 305.
524. Wittmann-Liebold, B., and Wittmann, H. G. (1967). Coat protein of two RNA viruses: Comparison of their amino acid sequences. *Mol. Gen. Genetics* **100**, 358.
525. Wood, W. B., Edgar, R. S., King, J., Lielausis, I., and Henninger, M. (1968). Bacteriophage assembly. *Federation Proc.* **27**, 1160.

526. Wu, R., and Kaiser, A. D. (1967). Mapping of the 5'-terminal nucleotides of the DNA of bacteriophage λ and related phages. *Proc. Natl. Acad. Sci. U.S.* **57**, 170.

527. Wu, R., and Kaiser, A. D. (1968). Structure and base sequence in the cohesive ends of λ DNA. *J. Mol. Biol.* **35**, 523.

528. Yamazaki, H., and Kaesberg, P. (1963). Degradation of bromegrass mosaic virus with calcium chloride and the isolation of its protein and and nucleic acid. *J. Mol. Biol.* **7**, 759.

529. Young, E. T., and Sinsheimer, R. L. (1964). Novel intra-cellular forms of λ DNA. *J. Mol. Biol.* **10**, 562.

530. Young, E. T., and Sinsheimer, R. L. (1967). Vegetative bacteriophage λ DNA. Infectivity in a spheroplast assay. *J. Mol. Biol.* **30**, 147.

531. Young, J. D., Benjamini, E., Shimizu, M., Leung, C. Y., and Feingold, B. F. (1963). Antigenic activity of tryptic peptides of tobacco mosaic virus protein. *Nature* **199**, 831.

532. Young, J. D., Benjamini, E., and Leung, C. Y. (1968). Immunochemical studies on the TMV protein. VIII. Solid-phase synthesis and immunological activity of peptides related to an antigenic area of TMV protein. *Biochemistry* **7**, 3113

533. Zinder, N. D., and Lederberg, J. (1952). Genetic exchange in *Salmonella*. *J. Bacteriol.* **64**, 679.

534. Zinder, N. D., Valentine, R. C., Roger, M., and Stoeckenius, W. (1963). f1, a rod-shaped male-specific bacteriophage that contains DNA. *Virology* **20**, 638.

535. Zinder, N. D., and Cooper, S. (1964). Host dependent mutants of the bacteriophage f2. *Virology* **23**, 152.

---

536. Arber, W., and Lin, S. (1969). DNA modification and restriction. *Ann. Rev. Biochem.* **38** (in press).

536a. August, J. T. (1969). Mechanism of synthesis of bacteriophage RNA. *Nature* **222**, 121.

536b. Bol, J. F., and Krauseman, J. (1969). The reversible dissociation of alfalfa mosaic virus. *Virology* **37**, 485.

537. Crawford, L. V., Follett, E. A. C., Burdon, M. G., and McGeoch, D. J. (1969). The DNA of a minute virus of mice. *J. Gen. Virol.* **4**, 37. Also (a) Crawford, L. V. (1969). The nucleic acids of the tumor viruses. *Advan. Virus Res.* **14**, 89.

538. Diener, T. O., and Raymer, W. B. (1969). Potato spindle tuber virus: A plant virus with properties of a free nucleic acid I. Assay. *Virology* **37**, 351.

539. Franker, C. K., and Gruca, M. (1969). Structural protein of the Friend virion. *Virology* **37**, 489.

540. Hariharasubramanian, V., and Siegel, A. (1969). Characterization of a new defective strain of TMV. *Virology* **37**, 203.

540a. Hung, P. P., and Overby, L. R. (1969). The reconstitution of infective bacteriophage Qβ. *Biochemistry* **8**, 820.

541. Kamen, R. (1969). Infectivity of bacteriophage R17 RNA after sequential removal of 3′ terminal nucleotides. *Nature* **221**, 321.
542. Kaper, J. M. (1969). Reversible dissociation of cucumber mosaic virus (Strain S). *Virology* **37**, 134.
543. Langenberg, W. G., and Schlegel, D. E. (1969). Localization of tobacco mosaic virus protein in tobacco leaf cells during the early stages of infection. *Virology* **37**, 86.
544. Lister, R. M., and Bracker, C. E. (1969). Defectiveness and dependence in three related strains of tobacco rattle virus. *Virology* **37**, 262.
545. Martin, M. A., and Axelrod, D. (1969). Polyoma virus gene activity during lytic infection and in transformed animal cells. *Science* **164**, 68.
545a. Min Jou, W., and Fiers, W. (1969). Studies on the bacteriophage MS2. VII. Structure determination of the longer polypurine sequences present in the pancreatic ribonuclease digest of the viral RNA. *J. Mol. Biol.* **40**, 187.
546. Ralph, R. K., and Wojcik, S. J. (1969). Double-stranded tobacco mosaic virus RNA. *Virology* **37**, 276.
546a. Riva, S., Polsinelli, M., and Falaschi, A. (1968). A new phage of *Bacillus subtilis* with infectious DNA having separable strands. *J. Mol. Biol.* **35**, 347.
547. Sauer, G., and Kidwai, J. R. (1968). The transcription of the SV40 genome in productively infected and transformed cells. *Proc. Natl. Acad. Sci.* **61**, 1256.
548. Sela, I., and Kaesberg, P. (1969). Cell-free synthesis of tobacco mosaic virus coat protein and its combination with ribonucleic acid to yield tobacco mosaic virus. *J. Virol.* **3**, 89.
549. Shatkin, A. J., and Sipe, J. D. (1969). RNA polymerase activity in purified reoviruses. *Proc. Natl. Acad. Sci.* **61**, 1462.
550. Strauss, J. H., Jr., Burge, B. W., and Darnell, J. E. (1969). Sindbis virus infection of chick and hamster cells: Synthesis of virus-specific proteins. *Virology* **37**, 367.
551. Verduin, B. J. M., and Bancroft, J. B. (1969). The infectivity of tobacco mosaic virus RNA in coat proteins from spherical viruses. *Virology* **37**, 501.
552. de Wachter, R., and Fiers, W. (1969). Sequences at the 5′-terminus of bacteriophage Qβ RNA. *Nature* **221**, 233.
553. Wimmer, E., and Reichmann, M. E. (1969). Two 3′-terminal sequences in satellite tobacco necrosis virus RNA. *Nature* **221**, 1122.
554. Wimmer, E., Chang, A. Y., Clark, J. M., Jr., and Reichmann, M. E. (1968). Sequence studies of satellite tobacco necrosis virus RNA: Isolation and characterization of a 5′-terminal trinucleotide. *J. Mol. Biol.* **38**, 59.
555. Yamazaki, H. (1969). Mechanism of early inhibition by an RNA phage of protein and RNA synthesis in infected cells. *Virology* **37**, 429.
556. Colby, C., and Duesberg, P. H. (1969). Double stranded RNA in vaccinia virus infected cells. *Nature* **222**, 940.
557. Subak-Sharpe, H., Timbury, M. C., and Williams, J. F. (1969). Rifamicin inhibits the growth of some mammalian viruses. *Nature* **222**, 341.

SUGGESTED READING*

It is impossible to read, remember, and evaluate critically and comparatively all the papers listed in the Bibliography, and the actual number of papers that should have been cited is easily ten times longer. I, therefore, have chosen quite arbitrarily about fifty papers that happen to stick in my mind. Some of them seem to me to exemplify significant advances, others to illustrate the furious pace of current progress and the urge to report minor steps quickly. Let it be assumed that I would grade these papers anywhere from A+ to C−, and thus none of my colleagues need feel slighted not to find any of their papers in this list.

August, 1969 (536a) (XI).
Campbell, 1964 (53) (XIII).
Capecchi, 1966 (56) (XI).
Crawford *et al.*, 1969 (537) (XII).
Crick *et al.*, 1961 (71) (IX).
de Wachter and Fiers, 1969 (552) (VI).
Duesberg, 1968 (88) (XII).
Feix *et al.*, 1968 (106a) (XI).
Fraenkel-Conrat and Singer, 1957 (125) (X).
Gierer and Schramm, 1956 (147) (VIII).
Goulian *et al.*, 1967 (153) (XI).
Harris and Knight, 1952 (167) (V).
Herrmann *et al.*, 1968 (180) (IX).
Hershey and Rotman, 1949 (182) (XII).
Hofschneider and Hausen, 1968 (189) (XI).
Josse *et al.*, 1961 (230) (VI).
Kaiser, 1957 (233) (XIII).
Kamen, 1969 (541) (VI).
Kornberg *et al.*, 1959 (257) (XI).
Lodish, 1968 (273) (XI).
Marmur *et al.*, 1963 (291) (VI).
Meselson *et al.*, 1957 (306) (VI).
Mills *et al.*, 1967, 1968 (309, 310) (XI).
Nathans *et al.*, 1962 (318) (XI).
Ptashne, 1967a, b (344, 345) (XI).
Reichmann, 1964 (351) (VIII).
Ritchie *et al.*, 1967 (354) (XI).
Roberts *et al.*, 1967 (355) (X).
Sänger, 1968 (381) (VIII).
Schuster and Schramm, 1958 (393) (IX).
Sela and Kaesberg, 1969 (548) (XI).
Shatkin, 1969 (404) (XI).
Siegel *et al.*, 1962 (413) (XI).
Singer and Fraenkel-Conrat, 1961 (419) (VI).
Singer and Fraenkel-Conrat, 1969 (421) (IX).
Sinsheimer, 1959 (424) (VI).
Spiegelman *et al.*, 1965 (428) (XI).
Stanley, 1936 (432) (I).
Streisinger *et al.*, 1964 (443) (XI).
Sugiyama *et al.*, 1967 (453) (XI).
Terzaghi *et al.*, 1966 (461) (IX).
van Vloten Doting *et al.*, 1968 (484) (VIII).
Vinograd *et al.*, 1965 (486) (VI).
Watson and Crick, 1953 (498) (VI).
Weissmann *et al.*, 1967 (508) (XI)
Winocour, 1969 (520) (XII).

* Roman numerals indicate corresponding chapters of this book and arabic numbers refer to bibliography.

QUESTIONS

Chapters I, II

1. What are the crucial differences between bacterial toxins, bacterial viruses, and bacteria?
2. Do all virus infections produce symptoms?
3. What are the two main types of response in infected organisms?
4. How many virus particles are needed to produce a lesion?

Chapter III

1. How are capsomere, capsid, virion defined?
2. The .... weight of tobacco mosaic virus is 40×10^6 daltons?
3. Which of the main methods used in purification of viruses focus on their particle weight? on their buoyant density? on their protein surface?
4. How can the number of virus particles per milliliter be determined?

Chapter IV

1. What are the general differences in the properties of viral proteins and cytoplasmic proteins?
2. Is protein denaturation reversible?
3. Which conditions favor protein, which nucleic acid isolation?
4. Which protein group reactions are frequently utilized in studies of viral conformation and degradation?
5. What are the differences, in principle and application, between ultracentrifugation in sucrose gradient, cesium chloride, and analytical ultracentrifugation?

Chapter V

1. What are the methods for selective splitting of peptide bonds?
2. How are terminal amino acids identified? peptides? acylpeptides?
3. How is disulfide crosslinking in proteins abolished?
4. Which features in the structure of the TMV coat protein offer functional advantages?
5. About how much of that structure has been found identical in the survey of several natural strains (or several hundred chemically produced mutants)?
6. What is the molecular weight range of viral coat proteins?

7. What general conclusions can be drawn from the observed variabilities in the amino acid sequences of the coat proteins of the TMV group? And those of the picorna phages?
8. Which is the smallest structural protein known?

Chapter VI

1. Which components of viruses are particularly sensitive to alkali? to acid? to detergents? (Give causes and mechanisms to account for these sensitivities.)
2. Which general features can be utilized in separating terminal from internal fragments of degraded RNA?
3. What is the specificity of pancreatic RNase and of T1 RNase?
4. RNA of much lower molecular weight than the virus equivalent (virus particle weight/nucleic acid fractional amount) is found in bromegrass mosaic virus, influenza virus, reovirus, cowpea mosaic virus, Rous sarcoma virus. Which of these nucleic acids fractions, if any, is infective? What is the significance of these polydisperse states of viral nucleic acid?
5. Which two mechanisms can produce DNA with cohesive ends?
6. What is terminal redundancy?
7. What is circular permutation?

Chapter VII

1. Which component(s) of influenza virus are responsible for antibody formation? What is the significance of type specificity and group specificity?
2. How is it possible to ascertain experimentally whether a certain viral component is virus-specific or not?
3. Which classes of viruses have complex and variable compositions, and for what reason?
4. What virus-specific enzymes occur in virions?

Chapter VIII

1. What determines the length of rod-shaped viruses?
2. What is the nature of covirus systems? of satellitism? of deficient and defective viruses?
3. What factors determine whether viral nucleic acid is infective upon isolation?
4. What experimental techniques have established the intrinsic infectivity of pure viral nucleic acids?
5. What are the biological functions of viral proteins?
6. What are top components and proviruses?

Chapter IX

1. What is the chemical basis of point mutagenesis?
2. What are the main advantageous features of nitrous acid? of hydroxylamine? of dimethyl sulfate as mutagens?
3. What are amber mutants?
4. Which is the key word in viral genetics: reconstitution, retribution, recombination, or redundancy?

Chapter X

1. Which type of particles can form spontaneously from virus protein alone: rods? icosahedral (spherical) particles? both? neither?
2. In what respect does reconstitution of isometric plant and bacterial viruses differ?
3. Which are valid criteria for successful reconstitution of a virus?

Chapters XI–XIII

1. What is the role of bacterial pili in infection?
2. Is DNA synthesis required in the course of replication of polio virus? polyoma virus? influenza virus? herpes virus? Rous sarcoma virus?
3. What are the structural requirements (circularity, double strandedness, etc.) for infectivity of øX174 DNA and of λ DNA?
4. What are the requirements for *in vitro* replication of phage RNA? øX174 DNA? of λ DNA? Which of these have been replicated *in vitro*?
5. In what manner can phage coat protein control the production of the different phage-specific proteins?
6. What are the main differences between lysogeny, transduction, and transformation?
7. What is the main difference between the Bryan high titer and the Schmidt Rupin strain of Rous sarcoma virus?
8. What is the nature of virus-specific products detectable in various types of transformed cells?

AUTHOR INDEX

Numbers in parentheses are reference numbers and indicate that an author's work is referred to although his name is not cited in the text. Numbers in italic show the page on which the complete reference is listed.

C

G

L

W

Y

Z

SUBJECT INDEX

I

L

M

W

X